Services and Business Process Reengineering

Series Editors

Nabendu Chaki, Department of Computer Science and Engineering,
University of Calcutta, Kolkata, India

Agostino Cortesi, DAIS, Ca' Foscari University, Venice, Italy

The book series aims at bringing together valuable and novel scientific contributions that address the critical issues of software services and business processes reengineering, providing innovative ideas, methodologies, technologies and platforms that have an impact in this diverse and fast-changing research community in academia and industry.

The areas to be covered are

- Service Design
- Deployment of Services on Cloud and Edge Computing Platform
- Web Services
- IoT Services
- Requirements Engineering for Software Services
- Privacy in Software Services
- Business Process Management
- Business Process Redesign
- Software Design and Process Autonomy
- Security as a Service
- IoT Services and Privacy
- Business Analytics and Autonomic Software Management
- Service Reengineering
- Business Applications and Service Planning
- Policy Based Software Development
- Software Analysis and Verification
- Enterprise Architecture

The series serves as a qualified repository for collecting and promoting state-of-the art research trends in the broad area of software services and business processes reengineering in the context of enterprise scenarios. The series will include monographs, edited volumes and selected proceedings.

More information about this series at http://www.springer.com/series/16135

Abhishek Rawat · Dipankar Deb · Jatin Upadhyay

Recent Trends In Peripheral Security Systems

 Springer

Abhishek Rawat
Electrical Engineering Department
Institute of Infrastructure Technology
Research and Management (IITRAM)
Ahmedabad, Gujarat, India

Dipankar Deb
Electrical Engineering Department
Institute of Infrastructure Technology
Research and Management (IITRAM)
Ahmedabad, Gujarat, India

Jatin Upadhyay
Electrical Engineering Department
Institute of Infrastructure Technology
Research and Management (IITRAM)
Ahmedabad, Gujarat, India

ISSN 2524-5503   ISSN 2524-5511   (electronic)
Services and Business Process Reengineering
ISBN 978-981-16-1207-7   ISBN 978-981-16-1205-3   (eBook)
https://doi.org/10.1007/978-981-16-1205-3

This Springer imprint is published by the registered company Springer Nature Singapore Pte Ltd.
The registered company address is: 152 Beach Road, #21-01/04 Gateway East, Singapore 189721,
Singapore

Dedicated to my wife Vidhi and my daughter Keshvi for their unconditional support and love.

Abhishek Rawat

Dedicated to my wife Indulekha and my son Rishabh for providing me unending love and support.

Dipankar Deb

This book is dedicated to Lord Shree Eklingji. I thank my parents Sushilaben and Vinodkumar, sister Dhruti, wife Krupa, and my daughter Heer for their overwhelming support and inspiration all the time.

Jatin Upadhyay

Preface

Since the past few decades, peripheral security has been receiving attention in the scientific world. Many technological developments have brought security arrangement issues to the fore. Of late, many organizations have begun to emphasize security systems not just for the key safekeeping-related matters but also to gain strategic advantages. This is because of information that helps in peripheral security and also realizes the associated objectives, and helps the security agencies in taking appropriate decisions. There is also the challenge of establishing appropriate peripheral security modules and procedures so that it can't be hacked or drilled easily in the structural context.

There are various kinds of organizations related to national security which require special measures to avoid possible acts of terrorism. Peripheral security is needed for expensive facilities and machinery that interest thieves. But apart from theft, we have to prevent access by unauthorized personnel to avoid accidents. This type of facility's security system is comparatively more complex and requires advanced technical equipment installed and maintained by qualified technical personnel of security systems. Each installation is different from the other, and so new developments in peripheral security help various organizations in diverse ways. This book presents some of such key technologies.

Railways are prime for connectivity and good transport, so there are several enhancements needed in the devices used for the security of the railway track. Due to many reasons like improper maintenance of tracks, inefficient track parts, improper switching of tracks, water runoff below tracks, etc., rail accidents are happening and are causing harm to humans as well as other resources. So a novel method for railway track protection is also covered in this book.

In Chap. 1, we have discussed different types of peripheral security modules, which are used nowadays, like CCTV cameras, facial and object detection-recognition, RF-based security modules, fingerprint-based controlled access, outdoor and indoor navigation techniques, etc. The strengths and weaknesses of these modules are also discussed in this chapter which may interest potential readers.

Most of the peripheral security systems have restrictions in recognizing persons, apart from an exorbitant execution cost. Therefore, in Chap. 2, we have discussed the data-rate-based peripheral security which is the new development. Through this system, identification and communication of the exact location and the number of unauthorized entries with the central command unit are possible at a reduced cost. In this methodology, we apply a certain distinct number of transmitters to authorize individuals, and the transmitter module transmits a signal of a fixed data rate and signal format. This signal of a specific data rate and signal format is detected and cross-verified to identify authorized entries. Data rates and signal format get updated at predefined periods from the Central Command unit for enhanced security. It has the competence to protect, observe, and control the remote areas from the central unit.

An avalanche is a mutual phenomenon that appears at high altitudes typically covered by snow. There is danger of being trapped, and many people lost their lives in avalanches because the rescue operation was not possible within the necessary time. In Chap. 3, we have focused on the system that is an outcome of a patent disclosure that gives precise monitoring of human beings/systems during an avalanche. The patent discloses a specific tracing method and increases the effectiveness in rescuing people stuck under snow by identifying their location more precisely than existing solutions.

The most common motive for installing security cameras is to defend the organization from unauthorized entry. But it has been seen that generally unsocial elements initially try to damage the camera or clean its memory to remove the evidence. Chapter 4 discusses a dual-purpose system that serves as a door closer and is equally useful for surveillance purposes. The currently deployed surveillance systems store a large amount of data, thereby consuming large memory spaces. The novel feature discussed here is object identification and classification for the desired area, along with controlled access provision. This design holds embedded devices within a hydraulic door closer, which is almost similar in size to a conventional door closer. Notably, it is tough to identify it to be any different from a regular automatic door stopper.

Chapter 5 discusses the outcomes of simultaneous localization and mapping algorithms from a 3D infrared vision sensor. It also discusses the limitation of vision-based and other sensory-based indoor navigation and object tracking methods. Received Signal Strength Index (RSSI) of a wireless node indicates how effectively the device can capture a signal from an access point through a value that effectively indicates whether one has enough signal strength. RSSI is a simple process of location estimation, and when available, no additional hardware is needed at individual network nodes. However, it isn't easy to estimate the position in an indoor environment directly from the RSSI values alone. Therefore, in this chapter, we have discussed a robotic system that analyzes the environment by capturing live images with RSSI. The optical character recognition algorithm identifies characters from images and stores those characters in a string. The algorithm matches the string with the database to get the location information. The results of the predicted position from the classification method are verified with the OCR algorithm. The robotic agent also takes the target input from the authorized person.

In the final chapter, we have proposed methods for failure mitigation of railway tracks using a predefined strain gauge module-based network to secure the railway network. This module's primary purpose is to analyze the real-time condition monitoring of railway tracks to prevent railway accidents due to track crack/deformation, caving of support strata underneath, and various other reasons. Failure may be because of the life cycle of the track, natural disasters, or some anti-social elements damaging the rail track. The discussed method can detect the instantaneous change in the track's response, record at various levels where we can read and analyze, and take corrective measures on the faulty portion of the railway track depending upon its severity level. Therefore, in this book, we have focused on specific expansions in peripheral securities, which we hope will change the security setup in the coming years and give the readers technical understanding of such developments.

The first two authors would like to acknowledge our student Mr. Urvish Prajapati's initial work in the form of a Master's Thesis with us on Peripheral Security Systems which formed the genesis for further work by the authors of this book. We acknowledge the support of Dr. Navneet Khanna and his staff at the Mechanical Engineering Workshop. We acknowledge Dr. Mahesh Mungule for his valuable inputs in conceptualizing the secured system for railway tracks. We also acknowledge the valuable time and efforts of our Laboratory assistants Mr. Sunny Kadia, Mr. Dhaval Joshi, and Mr. Mahesh Solanki who have contributed their facial images in many of the figures in this book.

Acknowledgement is given to the Springer Nature publication and Multidisciplinary Digital Publishing Institute (MDPI) for allowing us to use relevant material from the following papers:

- Upadhyay, J., Deb, D., Rawat, A.: Design of smart door closer system with image classification over WLAN. Wireless Personal Communications (Nov 2019).
- Upadhyay, J., Rawat, A., Deb, D., Muresan, V., Unguresan, M. L.: An RSSI-based localization, path planning and computer vision-based decision making robotic system. Electronics 9(8), 1326 (Aug 2020).
- Prajapati, U., Rawat, A., Deb, D.: Integrated peripheral security system for different areas based on exchange of specific data rates. Wireless Personal Communications 111(3), 1355–1366 (Nov 2019).
- Prajapati, U., Rawat, A., Deb, D.: A novel approach towards a low cost peripheral security system based on specific data rates. Wireless Personal Communications 99(4), 1625–1637 (Jan 2018).
- Surve, J., Mehta, V., Rawat, A., Kamaliya, K., Deb, D.: Low-cost 2 MHz transmitter for the detection of human trapped under the snow. In: Lecture Notes in Electrical Engineering, pp. 41–50. Springer Singapore (Nov 2019).

Ahmedabad, India   Abhishek Rawat
Ahmedabad, India   Dipankar Deb
Ahmedabad, India   Jatin Upadhyay
January 2021

Contents

About the Authors

Abhishek Rawat received his Bachelor of Engineering in Electronics and Communication Engineering (2001), from Rajiv Gandhi Technological University, Bhopal. He received his Master of Technology (2006) and Ph.D. (2012) from Maulana Azad National Institute of Technology Bhopal, India. He is currently an Assistant Professor at the Institute of Infrastructure Technology Research and Management (IITRAM), Ahmedabad, Gujarat, India. He has 16 years of research, academic and professional experience in different premier institutions. Dr. Rawat is also senior member IEEE and has published more than 50 articles in international journals, Book chapters, national and international conference proceedings. He received the Young Scientist Award from MPCOST Bhopal in 2007, involved in the field trails of the IRNSS receiver and published four Indian patents. His research interests include Navigation systems, Satellite Communication and Peripheral security etc.

Dipankar Deb completed his Ph.D. from the University of Virginia, Charlottesville, with Prof. Gang Tao, IEEE Fellow and Professor in the Department of ECE in 2007. He did his B.E. from NIT Karnataka, Surathkal (2000), and M.S. from the University of Florida, Gainesville (2004). In 2017, he was elected to be IEEE Senior Member. He has served as Lead Engineer at GE Global Research Bengaluru (2012–2015) and as Assistant Professor in EE, IIT Guwahati 2010–2012. Presently, he is a Full Professor in Electrical Engineering at the Institute of Infrastructure Technology Research and Management (IITRAM), Ahmedabad, Gujarat, India. He is Associate Editor of IEEE Access journal and Book Series Editor for Control System Series/CRC Press/Taylor and Francis Group. He is also a book series editor of "Studies in Infrastructure and Control" with Springer. He has published 36 SCI-indexed journals, 40 international conference papers, and 10 technical books with Springer and Elsevier. He holds 6 US Patents. His research interests include active flow control, adaptive control, cognitive robotics, and renewable energy systems (including wind, solar, and fuel cells).

Jatin Upadhyay received his M.E. in VLSI and Embedded Systems from Gujarat Technological University in 2015. He is currently a Ph.D. candidate at the Institute of Infrastructure Technology Research and Management, Ahmedabad, India. His current research interests include image processing, neural network, and cognitive robotics. He has worked on high-speed data transmission over the FPGA development board using a fiber optic data link.

Acronyms

AES	Advanced Encryption Standard
AI	Artificial Intelligence
ANN	Artificial Neural Network
ASK	Amplitude Shift Keying
BER	Bit Error Rate
BS	Base Station
CAGR	Compound Annual Growth Rate
CC	Central Command
CCTV	Closed Circuit Television Camera
CDF	Cumulative Distribution Function
CDRS	Cognitive Decision-making Robotic System
CRM	Central Receiving Module
CSV	Comma-Separated Values
DAQ	Data Acquisition
DATM	Data Acquisition and Transmission Module
DB	Database
D-MUX	De-multiplex
DoA	Direction of Arrival
DRM	Data Reception Module
FFT	Fast Fourier Transform
FPS	Frames Per Second
GPS	Global Positioning System
HF	High Frequency
IC	Integrated Circuit
ICTFF	Improved Color Texture Feature Fusion
IoT	Internet of Things
IR	Infrared
JPEG	Joint Photographic Experts Group
KNN	K-Nearest Neighbors
LC	Local Command
LIDAR	Light Detection and Ranging
MAC	Media Access Control Address

MAE	Mean Absolute Error
MCRM	Mobile Central Receiving Module
MISO	Multiple Input Single Output
ML	Machine Learning
MLP	Multilayer Perceptron
MOSFET	Metal Oxide Semiconductor Field Effect Transistor
MRM	Mobile Receiving Module
MSE	Mean Square Error
MUX	Multiplex
NMEA	National Marine Electronics Association
NN	Neural Network
NRZ	Non-Return to Zero
OCR	Optical Character Recognition
OOB	Out-Of-Bag
PA	Power Amplifier
PIR	Passive Infrared
PNG	Portable Network Graphics
RBF-SVM	Radial Basis Function kernel-Support Vector Machine
REST	Representational State Transfer
RF	Radio Frequency
RFC	Random Forest Classification
RGB	Red Green Blue
RM	Receiving Modules
RMSE	Root Mean Square Error
ROI	Region of Interest
RSA	Rivest-Shamir-Adleman
RSSI	Received Signal Strength Index
RT	Real Time
RTSP	Real Time Streaming Protocol
S/N	Signal-to-Noise Ratio
SDCS	Smart Door Closer System
SDR	Software-Defined Radio
SIFT	Scale-Invariant Feature Transform
SLAM	Simultaneous Localization and Mapping
SSH	Secure Shell Network
SVR	Support Vector Regression
TCP	Transmission Controlled Protocol
TM	Transmitting Modules
TOF	Time of Flight
UAV	Unmanned Aerial Vehicle
USFD	Ultrasonic Flaw Detection
UWB	Ultra-wideband
VHF	Very High Frequency
WLAN	Wireless Local Area Network

List of Figures

Chapter 1
Introduction

In recent times, the need for security in every walk of life has become inevitable. There are security cameras or other sensory systems installed in many places like railway stations, gated communities, shopping malls, schools, etc. The advent of surveillance cameras has brought about a revolution in the area of security. Nowadays, it has become imperative for cameras to be smart for various applications. The need for better security has encouraged researchers to work more upon the technological improvements in cameras where they not only capture the video but also analyze the captured content. A person's availability in the analysis of the captured video wherever an installed camera is available is highly impossible and encourages the surveillance systems with automatic video analysis. Thus, the research on automated surveillance systems is encouraged to a considerable extent. The shortcoming of the existing surveillance cameras is the lack of storage management. The emergence of facial recognition and image processing has brought about much development in safety and protection areas. The use of robotic surveillance systems adds a perfect edge to the research in the area, which encourages the use of the robotic surveillance systems in the commercial, public, and defense sectors.

To track or monitor an area remotely is the new area of investigation. The area covered with mountains, snow, or muddy water is difficult to analyze. The main challenge is to design a robust, low power consumption, environmental impact-free, and real-time system. Regions like mountainous areas covered with glaciers are the prime focus of our research work presented in this book. We explore different techniques for the localization of an object or person with stealth monitoring. The objectives are to identify methods for a robust solution for such problems. We identify techniques to sense the presence of an object in a specific region with the help of different radio frequency techniques and their limitations. On sensing an object's presence, we note the category and made a human level of cognition to observe the environment.

A. Rawat et al., *Recent Trends In Peripheral Security Systems*, Services and Business Process Reengineering, https://doi.org/10.1007/978-981-16-1205-3_1

1.1   Brief Review of Different Systems of Peripheral Security

An overall peripheral security system is a consolidated targeted localization approach that includes tracking, identification, and recognition. By sensing the changes in the signal quality or characteristics, an object gets localized. The GPS signal has its limitations of multi-path reflection in dense areas like the forest regions or indoors. Often, the GPS signals get purposely blocked due to many security reasons. In such cases, a low (compared to GPS)-range signal helps sense an individual's presence. A tracking device is a helpful tool that tracks an object/person in a critical situation like heavy snowfall. Image processing is a technique to identify a specific pattern, identify objects, and recognize it. In the subsequent sections, we present the significance of such methods and their suggested implementation as per the intended applications.

1.1.1   CCTV Camera

The usage of CCTV surveillance has become very widespread, leading to an increase in the number of stakeholders and users [1]. It has facilitated monitoring in both outdoor and indoor spaces. The footages obtained from CCTV camera surveillance systems demonstrate effective use in their application in legal investigations to prosecute criminals [2]. The use of surveillance has proven efficient in emergency conditions by providing immediate response through offense detection, thus reducing crimes. It has been suggested in the literature that most of the usage of surveillance is to prevent crime and thefts at public places [3]. A good quality capture of any object can benefit the viewer of the surveillance in many ways. Video surveillance has become a major part of highly crowded areas like railway stations, airports and supermarkets, offices, and residential areas to ensure maximum security [4].

The footages from conventional surveillance systems are often of not much use due to their low quality. The cameras cannot identify the critical facial features like the shape of the face, eye color, and hairstyle, which make up the essential facial characteristics. Conventional surveillance systems either have an operator who works from a control room to check on the activities or to record such actions in a digital format to undergo compression to save the storage space [5]. There is, thus, a steep decline in content quality, and it is difficult to identify the objects and people. Another essential factor that affects surveillance credibility is the positioning [6]. In such cases, the surveillance quality is also affected by the environmental conditions and may make the content futile. Due to these shortcomings, it is essential to incorporate some smart features for optimal usage. Some critical elements incorporated into the existing system are object detection, facial recognition, and tracking [7]. In the literature, we find that incorporating these features into the current system has proven useful in maximizing surveillance efficacy. The process of locating or identifying the objects present in a frame is called object detection.

1.1.2 Facial and Object Detection, and Recognition

Recognition of objects consisting of locating or identifying them in a single frame [8] is part of computer vision. Another aspect is object tracking that involves detecting moving objects even among multiple ones located in a single frame [9]. The object detection algorithms play a significant role in facial and object detection, which involves many factors such as the accessibility of high-power computers and a camera implementation area [10, 11]. The object detection algorithms employ a requirement of identification based on motion sensing, robotic surveillance, identification of moving objects, traffic surveillance, seamanship of vehicles, and many more [12]. We come across the plane and solid figures that are translated to 2D and 3D images for better understanding. While object detection is initiated, some parameters may disrupt the natural process that may be mislaying of information during communication, the appearance of noise during capturing, compounded motion of objects, compounded nature of objects, and many more [13]. It may not be possible to have the algorithms working without any disruptions, which contradicts the fact that there exists a plethora of image, object, and face recognition algorithms [14]. Some algorithms work based upon the characteristics of the image, including the size, shape, and modeling of the object involved in detecting the object and its tracking [15, 16]. The size and shape of the object voice the face of the object to be apprehended. The mundane object representation techniques consist of object representation, object detection, and tracking.

Representation of objects mainly conveys the entity's traits and characteristics distinguishable based on the entities' features and shapes. An object represented with the help of a point used for checking the object's path takes over an area of the conception produced by the surveillance. For 3D representations, the configuration of the shape is as per the shape model and articulation. The whole object is characterized in terms of its features while scanning its skeletal representation in articulation and density. The portrayal of the object constitutes the considerable share of extraction of characteristics. The next major step is to detect the changes in the object's position from frame to frame. A detection mechanism consists of a tracking algorithm to track down the changes in every frame. In the many detection mechanism algorithms that we come across, we can see that some use the data obtained from a single frame, whereas the more contemporary mechanism uses multiple frames. This facilitates minimum false detection. The types of real object detection methods are shown in Fig. 1.1.

Tracking of an object plays another major role in estimating the object's path of motion by keeping track of its site of existence, in which the frames are changing throughout the video [14]. The correlation between the frames throughout the captured video is significant. Object detection and object tracking procedures happen either together or simultaneously, one after the other. The embedded unit processing capability also plays a vital role in system performance [17]. When the process takes place simultaneously, the object detection mechanism first detects the objects, and later on, the tracking mechanism forms a correlation between the sequence of objects

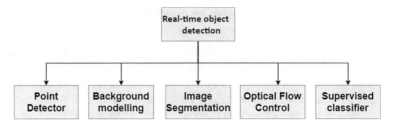

Fig. 1.1 Object detection techniques

located in the frame. If both the mechanisms function together, the object's spatial existence and the correlation take place coordinatedly. The estimation is done based on the data received from the captures.

1.1.3 Radio Frequency-Based Security System

The RFID technology is used to transfer the information through radio frequencies. This technology has the advantages of operating in a robust environment with low operating power. The line of sight connection is not needed to communicate between the transmitter and the receiver. The RFID technology is categorized based on the application, operating frequency, and energy sources. The main component of the RFID communication platform is the transponder and the receiver. The receiver are mainly RFID tags and these tags are categorized based on the relation to the energy. Active, passive, and semi-passive are the tag types. The active tags contain the battery sources and provide the energy to the tag chip as well as the antenna. Normally, the active tags work as beacons having the range of 5–10 m.

The RFID transceiver and tags are the main components of this technology. The transceiver transmits a particular range of frequency, and the change in signal amplitude reflecting back from the tag is detected by the reader. Every passive tag contains a coil and a memory chip. Transceiver generated electromagnetic signal like a primary side winding of transformer as shown in Fig. 1.2 as a copper winding. This signal induces the electromagnetic force to power the integrated chipset of the tag. Based on the tag protocol binary values ones and zeros, the IC shorts the circuit. This will reflect the change in the amplitude of the transmitter signal. This change is measured by the microcontroller unit and generates the tag id code as shown in Fig. 1.3.

The data is received in a serial transmission, each RFID tag in our hardware is starting from number "0900". Every tag contains a 12-digit unique id number.

The RFID tags are categorized based on the power consumption, application operating range, types of tag memory, and tag construction. The tag classification is shown in Fig. 1.4. Power consumption of the tag is categorized by active, passive, and semi-passive tags. Active tags contain a battery unit within them. Because of

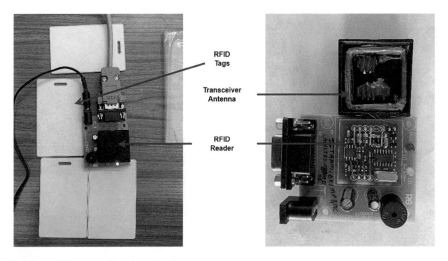

Fig. 1.2 RFID transceiver

Fig. 1.3 RFID serial communication with host computer

continuously transmitting pulses, it is also known as an RFID beacon. These tags are commonly used for applications like controlled access, supply chain management, measuring the timing of cars during a race, and office file location tracking.

The selection of RFID tags also depends on the application operating range [18]. Based on the range, the operating range of frequency varies from low-, mid-, and high-frequency bands. The lower band varies in between ≈124 and 135 kHz, mid range between ≈6.76 and 6.79 MHz, and high range from ≈13.55 to 13.56 MHz. The ultrahigh band and microwave frequencies are also used in RFID technology [19]. High frequencies are mainly used in near-field communication for applications like credit cards for contactless payments [20]. The tags are classified based on the memory type that is R/W (Read/Write) memory, Read-Only Memory, and Write Once Read Many (WORM). As discussed earlier, the selection of tags mainly depends on the application and usage. For example, many ink manufacturers of writing labels on packaging units want that printer and ink only be purchased from them. So the ink labels are encoded with the specific RFID code and printing count. The printer will not accept any after-market ink cartridges and the company gets a notification to send

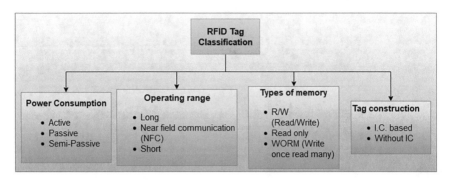

Fig. 1.4 Types of RFID tags

a new cartridge after several specific counts of printing. These types of tags are read and write-type tags. The write once read many types are tags used in supply chain management so that each type of product gets individual IDs. Read-only memory-type cards are the cheapest among all types of tags. Tags having the specific code and the codes are identified by the specific reader. The programmer writes a code that if this number of serial numbers is detected, then do as directed.

The tag construction can be done with and without integrated circuit based [21]. IC-based tags are normally powered from the battery (in case of active tags) or from the antenna after inducing energy after interaction with the reader signal. The IC contains the encoding and decoding as well as demodulation-modulation blocks. These blocks make secure communication between the transmitter and the receiver. They also contain the memory unit that is discussed in the previous paragraph. An efficient IC needs low power usage so it requires low power to operate; this utilized power can be used to extend the range of the tag. The passive tags without IC is an emerging topic of research in recent times [22–25]. The chipless tags are functioning based on system implementation. The detailed categorical information of an IC-less RFID tag is shown in Fig. 1.5.

This tag is characterized based on time domain [26], frequency domain, or a hybrid formulation. The time domain-based tags are further divided based on the modulation techniques such as Pulse Position Modulation (PPM), Phase Modulation, On-Off shift Keying (OOK), and Time Domain Reflectometry (TDR). In a hybrid tag, a single resonant substrate has more than two logic stages which are achieved with the help of frequency encoding. The RFID technology is used to identify any object's authenticity and location based on tag information without a line of sight interaction. The technology is further modified, and measuring a change in a data-rate-based peripheral security-based system is discussed in Chap. 2.

Fig. 1.5 RFID tags without IC

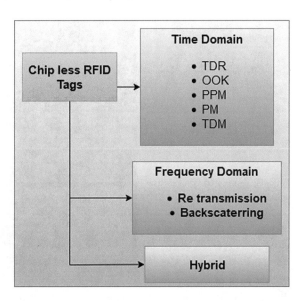

1.1.4   Fingerprint-Based Controlled Access

A fingerprint sensor is one of the biometric sensors among face, hand geometry, iris, retinal scan, voice, and signature [27, 28]. As the fingerprint of every person and even each finger is unique, it is commonly used for access control and to check the authenticity [29, 30]. The fingerprint-based module for the controlled area is shown in Fig. 1.6. A fingerprint sensor consists of a scanner unit, driver board or embedded board, and power supply unit for supplying power to the microprocessor and the scanner. The inputs are collected through the serial transmission with the help of an RS232 connector. The scanner unit captures the light that is reflected from the top of the exposed area that is visible at the glass. The reflected signals are collected by the Charge-Coupled Device (CCD) sensors [31]. The sensor is interfaced with the microcontroller unit to transfer the digital value of an individual finger to the processing unit through a serial connection. The glass cover is used because of its property like rigidness, high durability, and low sensitivity for electro-static discharges.

The performance of the device is calculated based on its positive detection, false detection, false acceptance rate, and false rejection rate. The average true positive detection rate is greater than 95%, false detection rate is less than 5%, false acceptance rate is less than 1%, and false rejection rate is less than 8%. But the false verification rate also depends on the programming software [32], that is, how much change in value is accepted or the percentage value that matches with the original image. If the value is not matching at a certain level, a device can ask the user for a re-scan.

Fig. 1.6 Fingerprint sensor
with embedded module

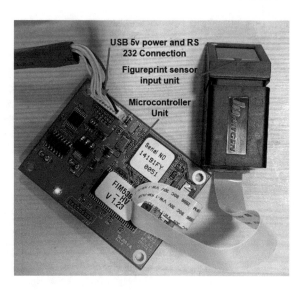

1.2   Current Studies in Computer Vision and Radio Signal-Based Indoor Navigation Techniques for Tracking

The conception of the latest technologies into the hands of laymen has brought about a swift revolution in recent times, which has made it so easy for us to control the devices around us using our Smartphones and wireless devices. Such technologies have emerged in many environments, majorly focusing most of their attention on indoor localization. Indoor localization facilitates the acquiring of the position of an object or a system in an indoor environment. This type of methodology needs to be employed to ease the surveillance concerns which may be faced. Indoor localization finds wide application in wireless communication, wireless sensor networks, industrial settings, and most major robotics. Even though not a very long time has passed since the conception of smartphones and wireless devices, their large-scale existence has made daily tasks time-efficient and labor-efficient by increasing the ease, accuracy, and decreasing the time taken to perform these tasks. The process of tracking down the user or the device has shown promising results in surveillance, disaster management, health sector, and industry by encouraging the conception of novel technologies like Smart Grids and Internet of Things (IoT). Such technologies promise to provide far and broad connectivity to a plenty of devices. The IoT technologies ensure the availability of information at low power consumption, but overlook the fact that there is a requirement for a high data rate at the indoor level. This requires numerous communication interfaces for both short range and long range to make the communication smooth and flawless. Indoor navigation of the smartphones and wireless devices is employed to the long range—IoT Technologies to figure out an indoor object's location [33].

The other technology that has been very widely used and plays a significant role in our daily lives is GPS. GPS provides us with the point of the existence of any device in the outdoor environment. In the indoor environment, GPS technology to locate any object does not show the same efficacy as in the outdoor environments. The basis of the failure may be the occurrence of obstruction in the signal, which may be caused by structures, walls, and many other objects present in the indoor environment. A WLAN, UWB, and RFID technology can be used instead of a GPS due to the lack of line of sight connection [34]. The non-performance of GPS in the environment has led to novel technologies that can locate specific indoor environments. These specially designed methods have been modified to suit the indoor locale called the indoor positioning systems.

1.2.1 Indoor Navigation Technologies

The technologies employed for Indoor Navigation are majorly classified into eight sensory techniques as shown in Fig. 1.7:

• Wi-Fi: The elemental technology which possesses the efficiency of networking and equipping the devices with an Internet connection to different gadgets and devices in the commercial, private, and public environments is called Wi-Fi. It is an IEEE 802.11 standard functioning and providing for the ISM band. At its inception, the expanse of Wi-Fi was around 100 m, which has now extended to up to 1 km. In our daily life, the gadgets used by us, ranging from Laptops, smartphone, TVs, and many more, are connected through a Wi-Fi network, making our work

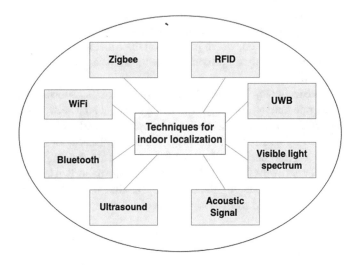

Fig. 1.7 Techniques used for indoor object localization

less laborious and easy. These advantages have made this technology a primary choice for applications of indoor localization. It has been reported in the literature that this technology has shown promising results due to the ease of the access points being utilized as the reference points for accumulating signals. Another major advantage provided by the Wi-Fi technology is the non-involvement of other added infrastructure in the space. The extant Wi-Fi technology is generally employed to ease the data transmission and keep a check on network coverage for smooth operation. Using efficient and smart algorithms, we can achieve higher object localization accuracy with a pre-installed wireless network.

- Bluetooth: When an expanse is considered, which consists of remotely located wireless devices that may be stagnant or in motion, they can be connected through Bluetooth. It is an IEEE 802.15.1 standard very widely used for easing the connection between wireless devices. The older versions of Bluetooth have not shown excellent area coverage, but the most recent version of Bluetooth, also called Bluetooth Low Energy or Bluetooth Smart, is shown to equip the systems with a data rate of 24 Mbps with the stretch of area that can be covered being 70–100 m radius. Bluetooth Low Energy shows favorable responses when employed with different localization techniques such as ToF, AoA, and RSSI, dependent mostly on the RSSI inputs [35]. Bluetooth Low Energy has advantages: providing a good range at a low cost and consuming less power [36]. The systems also have significantly less complexity, which increases the ease of working with these technologies. There are some Bluetooth Low Energy-based protocols introduced by Apple Inc. and Google Inc. named as iBeacons and Eddystone, which are majorly applicable for proximity detection by transmitting the signals for proximity detection at some specific time intervals. A 16-Byte UUID (Universally Unique Identifier) is given out as a Beacon message, and any Bluetooth Low Energy-enabled device can connect with it by using an optional 2-byte code, which consists of a Major value and a Minor value. When a device obtains the message from iBeacon [37], it can connect to the device with the help of a cloud to send or receive information. The technology of iBeacon, which is based on the Bluetooth Low Energy model, need not only be used only for proximity detection but can also be used for indoor localization [38].
- Zigbee: Zigbee technology has the advantages of low cost, low power consumption, and less data rate making it energy-efficient in personal area networks [39, 40]. It is an IEEE. 802.15.4 standard. Zigbee can be used as an indoor localization based on the received signal strength index [41]. The protocol stack consists of multiple layers where ZigBee occupies the stack's higher levels, making it efficient for wireless sensor networks [42, 43]. The other layers are majorly used for spreading communication and organizing the network. The drawback of the technology is that it cannot be accessed from most user devices, leaving it unsuitable for indoor localization.
- Radio-Frequency Identification (RFID): A Radio Frequency flexible circuit that is on the receiving end to accumulate the signals given out by an electromagnetic transmitter, which functions by transferring and storing the data, is called the RFID. The basis of the RFID transmission is a reader who can scrutinize the information

and communicate with radio frequency-enabled badges. The RFID badges can give out the data, and the reader can acquire the data established on a protocol fed into the RFID tag and the reader. The RFID technologies are majorly divided into two types based on the technology used: Active and Passive RFID as discussed in Sect. 1.1.3. The RFID technology has been used for indoor navigation in which the tags are placed at the fixed locations. The robotic agent tries to scan the environment based on the detected tag information, to identify the current position of the robot. The main drawback to using this system is that many times the tags are not detected by the scanner due to other environmental noise. Also, multiple tags are needed to install at a specific range; this makes the application more complex also the multi-tag detection can misguide the detection system. The use of active RFID can give very high coverage and work in microwave frequency and ultra-high-frequency ranges. In Active RFID, the reader is situated hundreds of meters away from the ID, powered by a nearby power source. The downside of the active RFID is that it is large, expensive, heavy, and adds noise to the environment. Passive RFID can be used only where a very less range of area needs to be covered, which may range 1–2 m. There is no need for any local power source, and it can operate on a battery. The range and installation complexity are the issues for a passive RFID-based indoor localization technique.

- Ultra-Wideband (UWB): This technology consists of short pulses transmitted with a large bandwidth of greater than 500MHz with the frequencies 3.1 and 10.6 GHz [44]. The pulses are ultra-short with a time period of less than one nanosecond and possess a very short duty cycle [45]. The low duty cycle results in low power consumption, an added advantage for communication systems used in indoor localization applications. It is also insusceptible to any disturbance from other signals emitted from other devices present in the locale. The frequencies can also pass through different materials like walls, furniture, and other materials except for metals and liquids, which may disrupt the signal flow [46]. The technology can be used in the indoor environment with some exceptions.

- Visible light communication: This technology can be used to transfer high-speed data within the bandwidth of 400 to 800THz. It is visible in the abovementioned bandwidth, which is effused by Light Emitting Diodes [47]. It has a wide-scale conception, even when compared to other technologies. The phenomenon behind this technology's communication is the use of light sensors that calculate the location of the emitter of LED by estimating the direction and distance [48]. The drawback in this technology is that the LED and the sensor must be located in the same room to provide the line of sight, which provides the object's accurate location [49]. Else, the functioning of the technology will be compromised.

- Acoustic Signal: Acoustics majorly means sound, which means there needs to be an employee of microphone sensors that absorb the sound signals from the emitter to evaluate the user's location and distance by the process of estimation [50]. The acoustic signal-based estimation of the location has proven to be accurate in position estimation. The frequency of signals is in the audible band, which is less than 20kHz in whose range the relative position can be accurately estimated. The method consists of transmitting acoustic signals transmitted from the sound sources consisting of all the related information. The Doppler effect principle has also been used to recognize the phase shift of the received signals to estimate the device's position whose location needs to be found [51]. The sound signals are low and do not cause any sound pollution/noise, which may require high-level signal processing algorithms to enhance the signal's power detected by the receiver. The drawback of this technology is the requirement of additional devices to amplify the signals, which may compromise the size of the circuitry.

- Ultrasound: The frequencies which come under the category of ultrasound are greater than 20 kHz. This technology is majorly based upon calculating the distance between the sound transmitter and the object whose location needs to be traced [52]. The accuracy achieved by the technology is high up to a centimeter-level while tracking multiple objects simultaneously with lower power consumption [53]. An RF pulse aids the synchronization between the ultrasound signal transmission. The RF pulses are affected by the velocity of sound, humidity, and temperature changes of the environment, encouraging the use of temperature sensors in the circuitry, increasing its size and cost. If there is a permanent source of noise in the existing frequencies, the system's performance is further decreased, which may prove it to be inefficient. These drawbacks do not encourage the use of this technology in wireless communication.

1.2.2   Challenges in Indoor Navigation and Object Tracking

Next, we illustrate the various challenges faced by localization techniques.

- Effect of multiple paths and noise: The signals undergo many physical phenomena based on their nature of reflection, refraction, and diffraction when projected on different materials like concrete, wood, metals, and also human skin in some cases. Such physical phenomena have been noted to show considerable deterioration in the performance of the localization technique. Many localization algorithms rely on the signal obtained from the user device to evaluate the user's location. Due to the signal's physical phenomenon, there is diversification of signals leading to multiple paths. The signal leads to multiple paths making it difficult to recognize a single signal carrying the source's information. The multiple signals obtained may be phase delayed, time delayed, or attenuated signals losing the signal's original fervor. This leads to distorted results that do not convey the original information of the signal. The accuracy of localization is affected due to this challenge faced by the signal. To overcome this drawback, highly complex signal processing techniques need to be employed to mitigate different paths' multiple signals. There are a significant number of noise suppression algorithms that have been introduced lately. The algorithms are very complex and require very high processing power, making it heavy in the localization algorithms. In applications with limited power availability and a high presence of obstacles, algorithms will require optimizers or filters to suppress the noise and multiple signals' multiple paths to enhance localization accuracy.

- Effect of radio signal interference: The indoor environment plays a significant role in the process of indoor localization. The localization system's efficiency is positively impacted by the materials used in the components that make up the environment. The existing algorithms do not work efficiently in the real-time environment because their testing is carried out in controlled environments that consist of characteristics that are far different from those in real-time environments. The assumption made in the controlled environments is that there is a minimum of at least one line of sight between the user and the RN. Nevertheless, in real-time environments that may be a house, apartment, mall, or an office, it is highly impossible for a single continuous path for the line of sight. It is also observed that the environment might not necessarily have the same activity throughout the day or year. For example, supermarkets are highly crowded during peak hours and may not be very crowded during the rest of the day.

- Efficiency: The most significant challenge the localization systems face is that the systems need to be highly energy efficient. They need to provide high localization accuracy with good scalability without discharging the battery fast and being energy efficient. The localization algorithm's signals need to be received regularly to monitor the signals actively throughout the localization procedure. The consumption of energy used by the device needs to be optimized to reduce the battery discharge time. Many localization algorithms have thrown light on improving localization accuracy, but less focus has been thrown upon energy consumption by the system used for localization. If the algorithms' complexity is reduced and noise suppression is done effectively, the algorithms will consume less energy and power, considerably reducing energy consumption. If the power supplied is continuous, then the focus needs to be the response time, which plays a preeminent role in updating its location.

- Security and privacy concern: The willingness to share location-related data has become obsolete as many users find it compromising their privacy and security. The many localization systems present do not pay heed to the concern related to privacy and security. Moreover, the indoor perspective localization algorithm needs to acknowledge privacy and security concerns, giving high priority to them compared to effective indoor localization. As far as privacy issues are concerned, the user needs to be assured that the data relating to the location is confidential and secure without being used for any marketing strategies. The service provider must also be liable for the privacy-related issue related to the data related to user location. Such issues are not in the algorithm's creator's reach, making it essential to tackle without fail. The peak increases in the malicious attacks have made the employment of localization algorithms complicated for the users. Validation of any new user is imperative for protecting the privacy and optimization of security. If any planned attack may be made, leading the system to the disclosure of the private details, it may be impossible to work to perfection in this case. However, it can be made possible to make both ends meet. The above discussion clears the issue that needs to be tackled regarding privacy and localization security, making it the most difficult challenge.

- Sensor cost: To include the indoor localization systems, some additional equipment may be required to overcome the discrepancies offered by the other parameters. For large reliability, appropriate software, required services, and database need to be arranged. The employment of these additional equipment leads to a considerable increase in the cost, which may be a major drawback for the user to choose the localization technologies. Apart from the already mentioned equipment, it is vital to have a cellular network and Wi-Fi nowadays in many indoor environments. Thus, additional equipment plays a crucial role in deciding the localization algorithm's factor.

1.3 Scope of Mitigation in Security Systems

Implemented security systems ensure the enhancement of security levels. The system must process the necessary information and pass critical information to the authority. The security of raw and processed data can directly be related to individual privacy. If any individual presence is sensed in a restricted area, that should be notified to the authority. But the person who is not present in the monitoring area and data related to that individual should not be shared with others. The monitoring devices like CCTV units, fingerprint sensor, RFID sensors contain sensitive information of individuals. This sensitive information is shared to use in authentic use terms. This information will generate devastating conditions if it falls in the wrong hands. The security and privacy of data are playing a crucial role to implement an overall peripheral security system.

A CCTV system is connected with the processing unit through a network, and each CCTV unit consists of a unique Internet Protocol (IP) address. The IP address can be found in any other device that is connected to the same network through a wireless node or local area network. If the security levels are not set properly then the live image data frames can be seen remotely by an unauthorized person. After getting access to the surveillance unit, the intruder can manipulate the data by inferring the camera setting such as decreasing the frame per second rate, reducing picture quality, or delaying frame update rate before transmitting data.

An RFID system detects tag information and is based on the received data algorithm that takes the decisions. Nowadays, many RFID receivers are available that catch the information of the tag. This information can be cloned to other blank tags and further it can be used to manipulate the system. The devices available in the open market such as "HackRF" are used to track the communication frequency between the transmitter and receiver. The communication pulses are captured and recorded; after that, the same signal is applied to the transmitter. The signal processing device cannot differentiate or authenticate that the information received from the other side is from the tag or any other tracking device.

The facial-object detection and recognition algorithm has its own limitation such as the lighting condition or objects overlapping/masking. It is difficult to capture the features from the image of lower quality. To extract the features such as facial area, person's clothing, and object pattern, the detection algorithm depends on many parameters such as camera angle and position, background lighting, captured image quality, image transmission rate (between camera and processing device), practical received image quality at receiving end, and embedded device processing speed. The rate of processing number of fps is the actual algorithm detection rate from the raw images. So the image processing algorithm has challenges such as camera installation skills, establishing secure data transmission, and real-time image processing and feature extraction.

Meanwhile, the Radio Frequency (RF)-based system object localization and detection method has the challenge of working over the dynamic robust environment. The signal property may be affected when transmitted in the environment of high voltage power transmission line or indoor environment where the signal gets multi-path refection from different objects. Also, the signal pattern can be analyzed by the monitoring tools and open-source software like GNU radio. This software can apply the filter operation over the captured signal, and the main context can be analyzed. This type of system can make the signal information inferior without acknowledging the transmitter and receiving devices.

This study aims to analyze the current studies in the security system design and practical implementation challenges. The implemented methods show promising results in a controlled environment like laboratories. But the environmental effect in a dynamic environment adds noise that makes the system unstable. This chapter has discussed the different technologies, implementation criteria, challenges to technology implementation, and limitations. The significance of CCTV for surveillance and the need for facial detection and recognition algorithms and the RFID and Fingerprint system are discussed. The techniques of indoor navigation and tracking of an object are challenging tasks. Wireless node, RF, RFID, UHF, ultrasound, and acoustic signal-based indoor navigation are the novel research topics in which researchers perform potential research work in recent times. The RF-based secure data-rate transmission-reception-based object tracking method is discussed in Chap. 2. The transmitter broadcasts a signal pulse train having a specific data rate. The signal characteristics, such as signal power (in dB), vary on moving toward and away from the receiving point. This change is sensed by the receiver unit and based on that receiver module analyzes the linear displacement. The detailed description of system hardware implementation is discussed in Chap. 2.

References

1. Banu, V.C., Costea, I.M., Nemtanu, F.C., Badescu, I.: Intelligent video surveillance system. In: 2017 IEEE 23rd International Symposium for Design and Technology in Electronic Packaging (SIITME). IEEE, Oct (2017)

2. Anghelescu, P., Serbanescu, I., Ionita, S.: Surveillance system using IP camera and face-detection algorithm. In: Proceedings of the International Conference on Electronics, Computers and Artificial Intelligence - ECAI-2013. IEEE, Jun (2013)
3. Akshay Bharadwaj, K.H., Deepak, Ghanavanth, V., Harish Bharadwaj, R., Uma, R., Krishnamurthy, G.: Smart CCTV surveillance system for intrusion detection with live streaming. In: 2018 3rd IEEE International Conference on Recent Trends in Electronics, Information & Communication Technology (RTEICT). IEEE, May (2018)
4. Chumuang, N., Ketcham, M., Yingthawornsuk, T.: CCTV based surveillance system for railway station security. In: 2018 International Conference on Digital Arts, Media and Technology (ICDAMT). IEEE, Feb (2018)
5. Xu, Z., Wu, H.R.: Smart video surveillance system. In: 2010 IEEE International Conference on Industrial Technology. IEEE (2010)
6. Seo, H., Choi, J., Kim, H., Park, T., Kim, H.: Short paper: surveillance system with light sensor. In: 2014 IEEE World Forum on Internet of Things (WF-IoT). IEEE, Mar (2014)
7. Yimyam, W., Pinthong, T., Chumuang, N., Ketcham, M.: Face detection criminals through CCTV cameras. In: 2018 14th International Conference on Signal-Image Technology & Internet-Based Systems (SITIS). IEEE, Nov (2018)
8. Zhou, X., Gong, W., Fu, W., Du, F.: Application of deep learning in object detection. In: 2017 IEEE/ACIS 16th International Conference on Computer and Information Science (ICIS). IEEE, May (2017)
9. Chatrath, J., Gupta, P., Ahuja, P., Goel, A., Arora, S.M.: Real time human face detection and tracking. In: 2014 International Conference on Signal Processing and Integrated Networks (SPIN). IEEE, Feb (2014)
10. Aslam, A., Curry, E.: A survey on object detection for the internet of multimedia things (IoMT) using deep learning and event-based middleware: approaches, challenges, and future directions. Image Vis. Comput. **106**, 104095 (2021)
11. Nayak, R., Pati, U.C., Das, S.K.: A comprehensive review on deep learning-based methods for video anomaly detection. Image Vis. Comput. **106**, 104078 (2021)
12. Zhao, Z.Q., Zheng, P., Xu, S.T., Wu, X.: Object detection with deep learning: a review. IEEE Trans. Neural Netw. Learn. Syst. **30**(11), 3212–3232 (2019)
13. Lal, M., Kumar, K., Hussain, R., Maitlo, A., Ali, S., Shaikh, H.: Study of face recognition techniques: a survey. Int. J. Adv. Comput. Sci. Appl. **9**(6) (2018)
14. Khan, M.Z., Harous, S., Hassan, S.U., Khan, M.U.G., Iqbal, R., Mumtaz, S.: Deep unified model for face recognition based on convolution neural network and edge computing. IEEE Access **7**, 72622–72633 (2019)
15. Yuan, Z.: Face detection and recognition based on visual attention mechanism guidance model in unrestricted posture. Sci. Program. **2020**, 1–10 (2020)
16. Garg, D., Goel, P., Pandya, S., Ganatra, A., Kotecha, K.: A deep learning approach for face detection using YOLO. In: 2018 IEEE Punecon. IEEE, Nov (2018)
17. Zhao, H., Liang, X.J., Yang, P.: Research on face recognition based on embedded system. Math. Probl. Eng. **2013**, 1–6 (2013)
18. Wu, D.L., Ng, W.W.Y., Yeung, D.S., Ding, H.L.: A brief survey on current RFID applications. In: 2009 International Conference on Machine Learning and Cybernetics. IEEE, Jul (2009)
19. Marrocco, G.: The art of UHF RFID antenna design: impedance-matching and size-reduction techniques. IEEE Antennas Propag. Mag. **50**(1), 66–79 (2008)
20. Chen, X., Yeoh, W.G., Choi, Y.B., Li, H., Singh, R.: A 2.45-GHz near-field RFID system with passive on-chip antenna tags. IEEE Trans. Microw. Theory Tech. **56**(6), 1397–1404 (2008)
21. Tedjini, S., Perret, E.: Radio-frequency identification systems and advances in tag design. URSI Radio Sci. Bull. **2009**(331), 9–20 (2009)
22. Herrojo, C., Paredes, F., Mata-Contreras, J., Martín, F.: Chipless-RFID: a review and recent developments. Sensors **19**(15), 3385 (2019)
23. Nair, R., Barahona, M., Betancourt, D., Schmidt, G., Bellmann, M., Hoft, D., Plettemeier, D., Hubler, A., Ellinger, F.: A fully printed passive chipless RFID tag for low-cost mass production. In: The 8th European Conference on Antennas and Propagation (EuCAP 2014). IEEE, Apr (2014)

24. Karmakar, N., Amin, E., Saha, J.: Chipless RFID Sensors. Wiley (2016)
25. Das, R.: Chip versus chipless for RFID applications. In: Proceedings of the 2005 Joint Conference on Smart Objects and Ambient Intelligence Innovative Context-aware Services: Usages and Technologies - sOc-EUSAI'05. ACM Press (2005)
26. Forouzandeh, M., Karmakar, N.C.: Chipless RFID tags and sensors: a review on time-domain techniques. Wirel. Power Transf. **2**(2), 62–77 (2015)
27. Zhang, D.D.: Automated Biometrics: Technologies and Systems, vol. 7. Springer Science & Business Media (2013)
28. Ashbourn, J.: Biometrics: Advanced Identity Verification. Springer, London (2000)
29. Derawi, M.O., Yang, B., Busch, C.: Fingerprint recognition with embedded cameras on mobile phones. In: Security and Privacy in Mobile Information and Communication Systems, pp. 136–147. Springer, Berlin, Heidelberg (2012)
30. Clarke, R.: Human identification in information systems. Inf. Technol. People **7**(4), 6–37 (1994)
31. Liu, M., Jiang, X., Kot, A.C.: Fingerprint reference-point detection. EURASIP J. Adv. Signal Process. **2005**(4) (2005)
32. Zabidi, S.A., Salami, M.J.E.: Design and development of intelligent fingerprint-based security system. In: Lecture Notes in Computer Science, pp. 312–318. Springer, Berlin, Heidelberg (2004)
33. Simões, W.C.S.S., Machado, G.S., Sales, A.M.A., de Lucena, M.M., Jazdi, N., de Lucena, V.F.: A review of technologies and techniques for indoor navigation systems for the visually impaired. Sensors **20**(14), 3935 (2020)
34. Kunhoth, J., Karkar, A., Al-Maadeed, S., Al-Ali, A.: Indoor positioning and wayfinding systems: a survey. Human-Centric Comput. Inf. Sci. **10**(1) (2020)
35. Uttraphan, C., Aziz, F.A., Wahab, M.H.A., Idrus, S.Z.S.: Bluetooth based indoor navigation system. IOP Conference Series: Materials Science and Engineering, vol. 917, p. 012055, Sep (2020)
36. Kalbandhe, A.A., Patil, S.C.: Indoor positioning system using bluetooth low energy. In: 2016 International Conference on Computing, Analytics and Security Trends (CAST). IEEE, Dec (2016)
37. Lee, C.K.M., Ip, C.M., Park, T., Chung, S.: A bluetooth location-based indoor positioning system for asset tracking in warehouse. In: 2019 IEEE International Conference on Industrial Engineering and Engineering Management (IEEM). IEEE, Dec (2019)
38. Satan, A.: Bluetooth-based indoor navigation mobile system. In: 2018 19th International Carpathian Control Conference (ICCC). IEEE, May (2018)
39. Larranaga, J., Muguira, L., Lopez-Garde, J.M., Vazquez, J.I.: An environment adaptive ZigBee-based indoor positioning algorithm. In: 2010 International Conference on Indoor Positioning and Indoor Navigation. IEEE, Sep (2010)
40. Luoh, L.: ZigBee-based intelligent indoor positioning system soft computing. Soft Comput. **18**(3), 443–456 (2013)
41. Goncalo, G., Helena, S.: Indoor location system using ZigBee technology. In: 2009 Third International Conference on Sensor Technologies and Applications. IEEE, Jun (2009)
42. Uradzinski, M., Guo, H., Liu, X., Yu, M.: Advanced indoor positioning using ZigBee wireless technology. Wirel. Pers. Commun. **97**(4), 6509–6518 (2017)
43. Dong, Z.Y., Xu, W.M., Zhuang, H.: Research on ZigBee indoor technology positioning based on RSSI. Procedia Comput. Sci. **154**, 424–429 (2019)
44. Zwirello, L., Schipper, T., Harter, M., Zwick, T.: UWB localization system for indoor applications: concept, realization and analysis. J. Electr. Comput. Eng. **2012**, 1–11 (2012)
45. Zafari, F., Gkelias, A., Leung, K.K.: A survey of indoor localization systems and technologies. IEEE Commun. Surv. Tutor. **21**(3), 2568–2599 (2019)
46. Alarifi, A., Al-Salman, A., Alsaleh, M., Alnafessah, A., Al-Hadhrami, S., Al-Ammar, M., Al-Khalifa, H.: Ultra wideband indoor positioning technologies: analysis and recent advances. Sensors **16**(5), 707 (2016)
47. Li, X., Yan, Z., Huang, L., Chen, S., Liu, M.: High-accuracy and real-time indoor positioning system based on visible light communication and mobile robot. Int. J. Opt. **2020**, 1–11 (2020)

48. Jerome, K., Tony, V., Vinayak, R., Dhanaraj, K.: Indoor navigation using visible light communication. In: 2014 Texas Instruments India Educators' Conference (TIIEC). IEEE (2014)
49. Afzalan, M., Jazizadeh, F.: Indoor positioning based on visible light communication. ACM Comput. Surv. **52**(2), 1–36 (2019)
50. Aloui, N., Raoof, K., Bouallegue, A., Letourneur, S., Zaibi, S.: Performance evaluation of an acoustic indoor localization system based on a fingerprinting technique. EURASIP J. Adv. Signal Process. **2014**(1) (2014)
51. Kapoor, R., Gardi, A., Sabatini, R.: Acoustic positioning and navigation system for GNSS denied/challenged environments. In: 2020 IEEE/ION Position, Location and Navigation Symposium (PLANS). IEEE, Apr (2020)
52. Li, J., Han, G., Zhu, C., Sun, G.: An indoor ultrasonic positioning system based on TOA for internet of things. Mobile Inf. Syst. **2016**, 1–10 (2016)
53. Medina, C., Segura, J., la Torre, Á.D.: Ultrasound indoor positioning system based on a low-power wireless sensor network providing sub-centimeter accuracy. Sensors **13**(3), 3501–3526 (2013)

Chapter 2
Design of a Peripheral Security Module Based on Exchange of Specific Data Rates

This chapter presents a practical approach to design-specific signal formats and data-rate-based peripheral security modules applicable for endorsement and acknowledgement functions for a particular set of persons. This module is useful in safety supervision and efficient observation of entrances of large peripheral zones. Mostly, conservative peripheral security modules are created based on RF modules. They rely on the power level or precise signal measurement, but those methods may easily be jammed or hacked. The proposed method focuses on identifying a legal person using the conversion of certain fixed data rates between transmitter and receiver modules. Experimental results of small peripheral module setup of hardware also cross-authenticate the projected method. The signal formats and data rates are reorganized at a systematic predefined time which permits additional security to the presently developed module. This method can be applied for activating extra alerts to protect unauthentic entry into large-scale zones which may be Defense camps, Civilian applications, etc.

The continuous observation of different locations of national importance of high risk is very vital. Many places take steps in security and need fast reaction like military units, prisons, nuclear power plants, etc. Presently, different types of security products are used in different places like a video surveillance system, active infrared perimeter security, leakage cable-type perimeter security systems, and electronic pulse-type perimeter security systems [1–5]. A security system based on video monitoring applies a camera shooting probe at the significant location of outer limits. The security person perceives the perimeter situation through a monitor to confirm safety. For immovable point monitoring, the camera shooting probe does not cover the entire periphery, and so the monitoring range is lesser. To cover large areas, many more cameras and spectators are required for the security module, making the entire system costly. Nishanthini et al. projected a smart video surveillance system which uses motion recognition to find the occurrence of an intruder [6].

Liu et al. built cutting-edge tracking systems to monitor moving humans across multi-camera using an improved color texture fusion algorithm to significantly improve the tracking accuracy in complex scenes involving changing speed and

© The Author(s), under exclusive license to Springer Nature Singapore Pte Ltd. 2021
A. Rawat et al., *Recent Trends In Peripheral Security Systems*, Services and Business Process Reengineering, https://doi.org/10.1007/978-981-16-1205-3_2

directions [7]. Another prevailing method applies an active infrared boundary protection system that emits infrared light, and the receiving end receives those rays. If this light gets cut by any person, then an IR receiver senses variation in light continuity and a microcontroller sends an alarm signal to the host [8]. The RSSI-based signal mapping and location-based method track the object inside the indoor environment. The RSSI signal link quality is dependent on the environmental parameters. The classification-based techniques help predict the position based on the signal properties [9]. This approach is applied in some places wherein a laser fence protects a large area.

A patented design for stealth monitoring at the sensitive area described by Deb et al. [10] is implemented within a hydraulic door closer unit. The monitoring device captures the motion of the image detected within the surveillance area. The captured frames are processed by the object and facial detection and recognition algorithm. The results of the algorithm are stored in the remote server [11]. Each type of security module/method has diverse controllability and price tag. High security and performance are at the cost of high price and increased system complexity. In this chapter, we propose a data-rate-based security system to overcome the limitations of the presently used methods.

2.1 Peripheral Security Design as per Data Rate

The data-rate exchange-based security approach can provide safety for large peripheral spaces and also legitimate information. In this system, we can directly find and interconnect precise locations and the number of illegal entries. The field strength of the modules determines the space between the RF transmitter and receiver. The proposed design has five subsystems: transmitter modules (T), receiver modules (Rm), sensor transmitter (S), sensor receiver, (R) and local command (LC) unit. The entire periphery of the secured area gets covered by n sensor transmitter and receiver pairs, as shown in Fig. 2.1.

Each sensor transmitter (S1, S2, S3, …, Sn) has its coverage area, and each sensor receiver (R1, R2, R3,…, Rn) has its coverage area in association with the sensor transmitter. These sensor devices are at certain distances below and above the ground based on the sensitivity and threat perception along the campus periphery. Once the system is installed, the sensor transmitter continuously transmits some specific signals and the sensor receiver receives those signals. These devices sense the presence of a person or any other object. Any discontinuity leads to the activation of Receiver Modules (RM) which come under the respective area of the Local Command (LC) unit so that predictive applications are not active at all times, to conserve energy [12]. The entire geographical area is classified in the predefined sub-sector. Each sub-sector has one receiver module which includes a receiver unit, a microcontroller, and an associated transceiver unit.

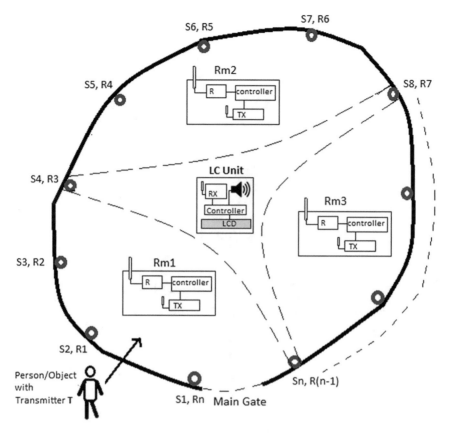

Fig. 2.1 Block diagram of proposed module (sensor transmitter and receiver represented by S and R, respectively)

The receiver modules set up in a particular area get linked with the LC unit through a transceiver T_x that gives the authenticity status. The number of transmitter modules (T) varies with the number of authorized persons allowed in that secured location. However, the number of receiver modules (R_m) is pre-decided as per the area. Each Transmitter module consists of a transmitter unit, microcontroller, and a receiver unit associated with the battery. One possibility is planting in the shoes or mobiles. The transmitter module uses the Amplitude Shift Keying (ASK) technique in which carrier signal strength is varied to represent a digital modulating signal, as given by

$$V_{ASK} = b(t) \, A \, Cos(\omega_c t), \tag{2.1}$$

where V_{ASK} is amplitude shift keying wave (Volt), A is the unmodulated carrier amplitude, and ω_c is the analog carrier radian frequency (radians/second).

In the above equation, $b(t) = 1$ for $V_m(t) = V$ and $b(t) = 0$ for $V_m(t) = -V$. Here, $V_m(t)$ is a normalized binary waveform to be transmitted. The carrier signal is

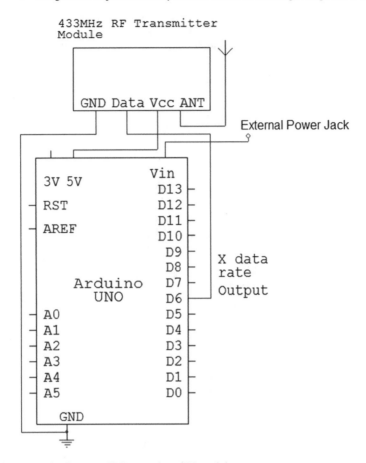

Fig. 2.2 Schematic diagram of RF transmitter (T_x) module

"on" when $Vm(t) = +V$, and it is off when $V_m(t) = -V$ [13]. Figures 2.2 and 2.3 show the circuit diagram of the initial transmitter and receiver modules.

The LC unit contains transceiver unit R_x, microcontroller, and alarm creating devices. The transceiver (R_x) connects with the transceiver (T_x) contained in the receiver module (R_m) and also communicates with the Transmitter module (T). The LC unit has a microcontroller connected with alarms, and a Buzzer operates when an unauthorized entry gets reported. The memory unit, when added to the LC unit, records the entries in a registry. Receiver modules and LC unit both self-check their operation. Therefore, when encountering physical damages, operational failure, or a sort of attack, a warning signal is generated for investigation.

We have used two transmitters and one receiver module operating at 433 MHz with one pair of IR sensors. A published patent details this invention [14].

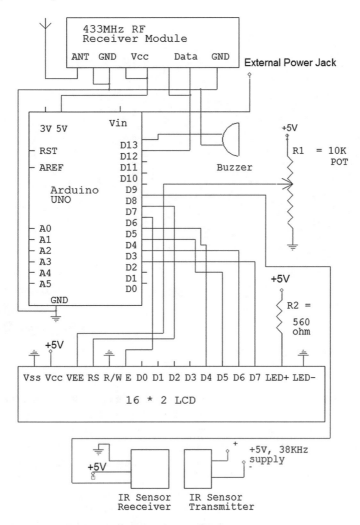

Fig. 2.3 Schematic diagram of RF receiver (R_x) module

2.2 Operational System Module

The predefined secured zone contains n number of transmitters. Each official person is fitted out with a small transmitter module. Sensor transmitters and sensor receivers are implanted in the feasible part of the unauthorized entry. The proposed system's architecture is shown in Fig. 2.4. The linkage between the transmitter module and the receiver module is well-defined as a data link. Receiver modules report to the LC module, which is shown as a reporting link. The wired link shows the sensor receiver's connection with the receiver module, which is connected using WSN.

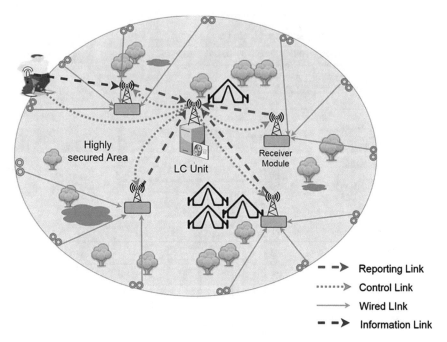

Fig. 2.4 Architecture of proposed system

Specific signal format and data rate are informed in receiver modules and transmitter modules by the LC unit at a precise time interval demarcated in a control link. The data rate is the total bits transported per second between a transmitting and a receiving section. We produce a precise signal of diverse data rate via encoding from the controller as per the obligation and the transmitter module's data-transfer capacity. In the present system, we applied a data rate of 500 bits/s. The transmitter module uninterruptedly transmits an RF signal with a predefined data rate and signal format. When an intruder comes into a protected area, the IR sensor receiver senses the object interruption and activates the receiver module. Then, the microcontroller located in a receiver module hunts for the specific data rate with the signal format.

Since an unauthorized person cannot transmit a specific signal of predefined data rate and so in the absence of the approved data rate, the receiver module directly reports this anomaly to LC with the location information. At LC, the alarm gets triggered, and the intruder's status and location information get stored in the registry of LC. When the approved person initiates the sensing module, the receiver is activated. The microcontroller identifies that the received data rate of a specific signal format matches a predefined data rate and sends information to LC about an authorized entry. An LC unit from time to time transmits specific coded signals with the data rate and signal format, to be defined at the receiver and transmitter modules for certification. The flowchart shown in Fig. 2.5 characterizes the discovery and verification process in a proposed module.

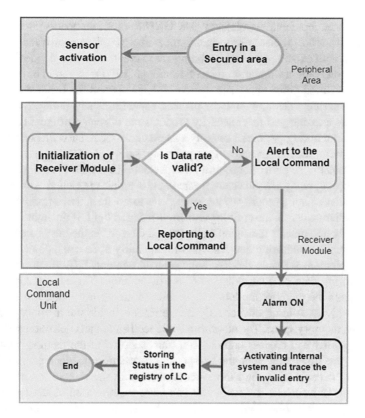

Fig. 2.5 Sequential flow graph of proposed system

2.3 Simulation Setup of Peripheral Security Using GNU Radio

Commonly, conservative security modules sense every person and report approval irrespective of inspection for legitimacy. This projected method identifies the certified people subject to the conversation of particular secured data rates among the transmitter and receiver systems. The data signal is commonly known as the baseband signal. These baseband signals change between two voltage levels, like +5 V for a binary 1 and −5 V for a binary 0. Non-Return to Zero (NRZ) is the furthermost used binary data format. The data rate is a rate of transmission of data signal commonly represented in bits per second (bits/s). The signal format indicates which specific signals are used as a data signal of the baseband signal to establish a reliable communication link.

Software-Defined Radio (SDR) is an auspicious radio technology used to implement a replica of communication scenario in software in place of hardware. The benefits of an SDR system include re-configuration and flexibility. But it suffers from low throughput and high latency. The GNU Radio is an open-source software enlargement toolkit. It also offers signal processing blocks to contrivance SDR for application with existing external RF hardware to create software-defined radios or without hardware in a simulation-like environment [15]. Observation of safety may be a state-of-mind perception. The degree of demands and the increasing budgets of security systems need novel and enhanced capabilities and innovative methodologies. The existing perimeter security products are of many different types, such as a video monitoring perimeter security system and a video surveillance perimeter security system, the leakage cable-type perimeter safety protection system, and the electronic pulse-type perimeter security and protection systems. If continuous scanning is adopted, many resources are wasted, and a monitoring blind area is easy to form. Another existing security system method is to use an active infrared perimeter security system wherein the emitting end emits infrared light. If the light is blocked, the receiving end detects the signal's change and then sends the host's alarm signal.

Small stealth monitoring devices with the capability of face and object detection can be implemented inside the hydraulic door closer unit [10]. Such imperceptible monitoring devices control access to restricted areas. The current surveillance system stores the memory module images after motion sensors detect the aperture area's sensitivity. Adding unnecessary image frames inside the memory unit only utilizes the memory space. The algorithm with facial and object detection and recognition capability can process unprocessed frames stored in the memory unit. This algorithm gets implemented on the low-power processing embedded devices. These devices are capable of storing the processes result in the remote server [11]. To track indoor movements, the RSSI and computer vision-based object localization algorithm provide a robotic agent location within the 2.4 m radial accuracy [9]. We propose a peripheral security system based on data-rate exchange to close loopholes of present methods and give the legitimacy status by sensing wrong entries.

GNU Radio is a freely available source toolkit for software development that provides signal processing blocks for software radio implementation and can be used in a simulation-like environment with low-cost external RF hardware. It is widely applicable to facilitate wireless communication science and real-world radio systems in amateur, academic, and industrial environments. The associated flow graphs are generally provided in C++ or Python. The signal source block produces a baseband signal of 100 Hz; square waves and other signal source blocks produce a sinusoidal carrier of 433 MHz. Signal amplitude levels get changed by re-configuring and adding a constant. The transmission channel model reproduces the signal behavior during transmission as an electromagnetic wave. To investigate a particular case when more than one intruder enters simultaneously and require validation. As the receiver moves away from the transmitter, signal power declines, so more noise will affect the signal. Along with multi-path effects, channel noise also added up with baseband signal. Mostly, source noises have large bandwidth and are additive in nature. The obtained signal on demodulation and the Bit Error Rate (BER) of the received signal verify acceptable noise level. This study follows the distribution of the receiver module throughout the geographical area. Data-rate-based peripheral protection frameworks are suggested in this chapter and validated using GNU radio-based modeling. Findings from this model also validate the idea. Received signal data rate and BER are acceptable at a prescribed area concerning the transmitters, and differences help authenticate the intruder. A specific way of providing security is data-rate-based peripheral security, enabling each intruder's authentication. This system is economical in cost and suitable to protect large peripheral areas in remote locations.

2.4   Results and Discussions

We observe the signal from the transmitter for the physical endorsement of the proposed approach. Initially, one transmitter module and receiver module are kept active at 433 MHz. Typically, the RF Time of Flight (TOF) and power-level measurement approaches help determine the distance between the involved wireless devices [16]. Measurement of power with low-power transmitters requires expensive apparatus and a detailed circuit design. After some range, even the change in power level is difficult to track, as shown in Fig. 2.6.

By keeping the transmitter at different locations, we measure the power using the N9936A Handheld Microwave Spectrum Analyzer [17]. We find a certain amount of changes in the power level for higher distances from the transmitter. Figure 2.6 indicates that beyond 10 m, the measured power level significantly changes, and the signal strength gets diminished. From the experimental data in Fig. 2.7, we observe minor changes in specific signal data rate up to some range, and then the variations in data rate are high and random, which may be due to various transmission impairments.

The presence of variations in data rate indicates the existence of transmitted signals within a specific proximity. Such variations in the received signal enable the

Fig. 2.6 Received signal power versus distance

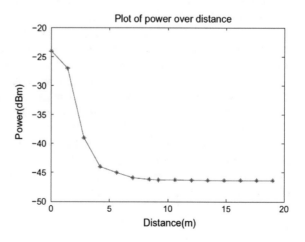

Fig. 2.7 Change in data rate
versus distance

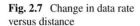
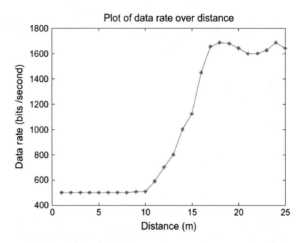

Fig. 2.8 T_x and R_x module
hardware setup

transmitter's tracking and authentication. We identify unauthorized entries by recognizing unauthorized data rate or absence of the same, as per Fig. 2.8.

We applied diverse links for communication at altered stages. This experiment is
undertaken in our laboratory (restricted to 10 m), but with an enhanced signal at the
transmission section, we can accommodate longer distances in real-life applications.
In a practical scenario, the region may differ in contrast to the experimental data.
Therefore, we find that distance at which data rate goes unchanged through experiments. The cost of the transmitter module with the current features is approximately
10 USD.

The peripheral security system is a low-cost security solution to detect a person's presence in the monitoring area. The sensors are activated after sensing the
movement in the monitoring area. The microcontroller unit compares the data rate
of transmission continuously. If an object or person impedes the transmitter and
receiving module path, it changes the signal pattern. After positive detection, the
controller unit informs the authority of the detection area location. The object or
the unknown person does not have the transmitter unit to validate authenticity. In

this case, the system identifies this event as a security breach and tries to trace the location range in which the object/person is detected. We use this RF signal-based object method to track the person's location under the snow. The system identifies the person's location by tracing the beacon's transmitter frequency. The transmitted signal has to pass through the thick sheet of snow and the multi-path reflection and refraction from the water particles. We can further increase the signal penetration by improving the transmitter power and frequency. During the implementation, the transmitter is placed under the heavy snow sheet, which replicates multilayered snow crystals. The received signal power and power spectral density are measured. The hardware design and schematics are further discussed in Chap. 3.

References

1. Wei, C., Yang, J., Zhu, W., Lv, J.: A design of alarm system for substation perimeter based on laser fence and wireless communication. In: 2010 International Conference on Computer Application and System Modeling (ICCASM 2010). IEEE, Oct (2010)
2. Lee, J.H., Kim, Y.S., Kim, B.K., Ohba, K., Kawata, H., Ohya, A., Yuta, S.: Security door system using human tracking method with laser range finders. In: 2007 International Conference on Mechatronics and Automation. IEEE, Aug (2007)
3. Jaafar, A., Kassim, M., Haroswati, C.K., Yahya, C.K.: Dynamic home automation security (DyHAS) alert system with laser interfaces on webpages and windows mobile using rasberry PI. In: 2016 7th IEEE Control and System Graduate Research Colloquium (ICSGRC). IEEE, Aug (2016)
4. Allwood, G., Wild, G., Hinckley, S.: Optical fiber sensors in physical intrusion detection systems: a review. IEEE Sens. J. **16**(14), 5497–5509 (2016)
5. Chunduru, V., Subramanian, N.: Perimeter-based high performance home security system. In: 2007 IEEE International Symposium on Consumer Electronics. IEEE, Jun (2007)
6. Nishanthini, S., Abinaya, M., Malathi, S.: Smart video surveillance system and alert with image capturing using android smart phones. In: 2014 International Conference on Circuits, Power and Computing Technologies [ICCPCT-2014]. IEEE, Mar (2014)
7. Liu, J., Hao, K., Ding, Y., Yang, S., Gao, L.: Moving human tracking across multi-camera based on artificial immune random forest and improved colour-texture feature fusion. Imaging Sci. J. **65**(4), 239–251 (2017)
8. Agarwal, N., Nayak, G.S.: Microcontroller based home security system with remote monitoring. IJCA Special Issue on International Conference on Electronic Design and Signal Processing ICEDSP (3), pp. 38–41, Feb (2013)
9. Upadhyay, J., Rawat, A., Deb, D., Muresan, V., Unguresan, M.L.: An RSSI-based localization, path planning and computer vision-based decision making robotic system. Electronics **9**(8), 1326 (2020)
10. Deb, D., Rawat, A., Khanna, N., Agrawal, S., Radadiya, H.: Smart imperceptible door closing systems with integrated monitoring (June, 30 2017), Indian Patent 201721021291 A
11. Upadhyay, J., Deb, D., Rawat, A.: Design of smart door closer system with image classification over WLAN. Wirel. Pers. Commun. **111**(3), 1941–1953 (2019)
12. Madni, A.M.: Smart configurable wireless sensors and actuators for industrial monitoring and control. In: 2008 3rd International Symposium on Communications, Control and Signal Processing. IEEE, Mar (2008)
13. Singh, R.P., Sapre, S.: Communication Systems: Analog and Digital, 3rd edn. Tata McGraw-Hill Education Pvt. Ltd (2012)

14. Rawat, A., Deb, D., Rawat, V., Dhaval, J.: Methods and systems for data rate based peripheral security (2017)
15. Truong, N.B., Suh, Y.J., Yu, C.: Latency analysis in GNU radio/USRP-based software radio platforms. In: MILCOM 2013 - 2013 IEEE Military Communications Conference. IEEE, Nov (2013)
16. Lanzisera, S., Zats, D., Pister, K.S.J.: Radio frequency time-of-flight distance measurement for low-cost wireless sensor localization. IEEE Sens. J. **11**(3), 837–845 (2011)
17. Keysight-Technologies: Fieldfox handheld analyzers (2018). https://assets.testequity.com/te1/Documents/pdf/keysight/FieldFox-TechnicalOverview.pdf

Chapter 3
System Design for the Detection of Humans Trapped in Snow

It has come to light recently that about a hundred Army personnel in India have lost their lives due to avalanches in the past 3 years. Such incidents highlight the challenges that the Forces face in providing relief measures and in remaining safe themselves. An avalanche is a common phenomenon faced in high-altitude areas typically covered by snow for most parts of the year. Such a situation poses fear of being trapped, and in fact, several lost their lives in an avalanche because the rescue operation was not possible within the necessary time.

When an avalanche strikes and the rescue team is still far away, there is a small chance to survive. During a full burial under snow with no visual signs available, transceivers may be useful in tracing the victim. In the last few years, many people have died due to an avalanche when the rescue operation could not happen effectively and swiftly. The proposed system is an outcome of a patent disclosure that gives precise monitoring of the location information of human beings/systems during an avalanche. In such cases, with relative ease, we can identify and locate the buried persons. A specific system is proposed for the tracing and increasing the effectiveness in rescuing people who are stuck under snow by identifying their location more precisely than existing solutions. The proposed design focused on the 2 MHz spectrum, which is suitable to penetrate the snow layer.

A human trapped under snow within 5 min has a 90% probability to be alive, but the chance to save the individual drops under 20% if 45 min have passed, and after 2 h, the chance is almost nil. Therefore, quick detection of the precise position of the victim under snow is critical [1]. The common methods for the rescue system include an approximation of the last known position and using sniffing dogs. The whole area can also be scanned by helicopters and search teams. The problem with this technique is its unreliability and the time that it takes to undertake the rescue operation. Looking for victims that are under the snow by manually scanning is not effective, and many sniffing dogs may be required depending on the area. The signal travels from one end to another from the dense snow, and the signal level

A. Rawat et al., *Recent Trends In Peripheral Security Systems*, Services and Business Process Reengineering, https://doi.org/10.1007/978-981-16-1205-3_3

characteristics may change after multiple reflections and loss of energy. Analysis of signal patterns reflected from the snow is a challenging task. A neural network-based classification method may detect the signal pattern, that is, transmitted by the beacon [2]. The data rate gets affected after reflection when a specific signal encoded with protocol bits is transmitted [3, 4]. This technique gets tested by the open-source simulators and hardware whose specific frequency range can penetrate the objects [5, 6]. The same problem is faced during the indoor localization, and the Received Signal Strength Index (RSSI) of multiple nodes are affected by the indoor environment. The classification-based techniques are implemented over the low-power embedded devices like Raspberry Pi 3B modules [7].

In the last few years, many rescue operations in the avalanche have been unsuccessful, resulting in death, and even there are cases when no dead bodies could be recovered. Manual scanning is ineffective when someone gets trapped deep inside the snow. Another technique for the detection used by professional hiking enthusiasts is the use of avalanche beacons. These beacons are transceivers, which always operate in the transmission mode. When any team member goes missing, the rescue team switches avalanche beacons in the receiving state and detects a possible signal from the beacons trapped in the snow. This rescue technique is much better than sniffing dogs, but it still has to cover the entire area manually, and to locate a trapped person/persons in this way is not pinpointed and exact. These methods are not real time. They are slow, and by the time the victim is located, it might be too late to save a life. If we can rescue the victims within half an hour of an avalanche, there are decent chances of survival. This requires a real-time system that pinpoints a victim's location on the map and should give their approximate depth so that digging out the victim can be planned accordingly. The proposed system can be applied to track its users all the time and when an avalanche hits.

3.1   Proposed System Design

The proposed system is shown in Fig. 3.1 and consists of six basic subsystems:

1. Receiving Modules (RM),
2. Transmitting Modules for specific person (TM),
3. Central Receiving Module (CRM),
4. Central Command (CC),
5. Mobile Receiving Module (MRM), and
6. Mobile Central Receiving Module (MCRM).

The methodology of the proposed system includes the identification of the location of Transmitting Modules (TM) with each user in an area around him. A predefined area is governed by the system, where a fixed number of Receiving Modules (RM) are installed. All these subsystems have a clock that is synchronized with each other. A specific range of frequencies can travel through the thick layers of the snow [8].

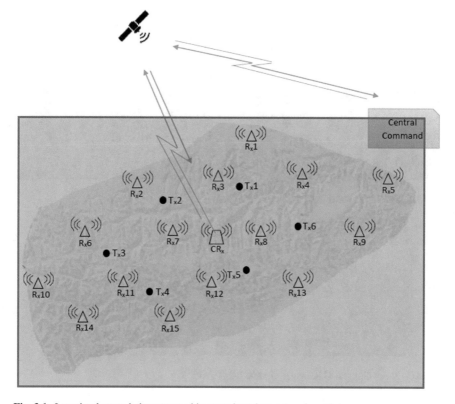

Fig. 3.1 Locating human beings trapped in snow based on network modules

The following section explains the methodology of selecting a range of frequency and hardware implementation.

The TM is primarily used to ascertain a user's location and is responsible for transmitting the status flag, approximating the depth at which the user is inside the snow. The signal gets transmitted at a frequency of around 2 MHz. This subsystem also contains a power amplifier bank to be used when the user is too deep inside the snow for obtaining a considerable signal strength at the Receiving Module (RM). Figure 3.2 describes the flow diagram of Transmission Module (TM).

The Receiving Modules (RM) shown in Fig. 3.3 receive signals from different TMs. Such signals get transmitted to the Central Command (CC). At the RMs, the signal gets filtered for noise. It would be highly unreliable to transfer such a signal from each RM to CC, and so a Central Receiving Module (CRM) is established.

Figure 3.4 describes the block diagram of the Central Receiving Module (CRM) unit. Each RM sends a status flag which indicates whether it's functioning or not, to the corresponding CRM.

The CRM in turn transfers that information to the Command Center (CC). In case any RM is destroyed by an avalanche, its status flag will change to "failure".

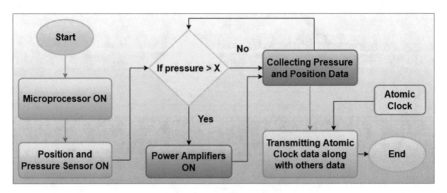

Fig. 3.2 Flow diagram of the Transmission Module (TM) unit

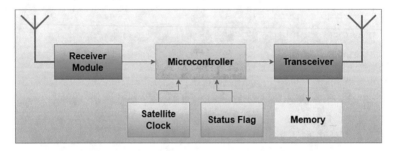

Fig. 3.3 Block diagram of Receiver Module (RM) unit

Then a Mobile Receiving Module (MRM) gets deployed at the location of the failed RM. This MRM is a mobile version of RM and can handle data transfer during such emergencies.

The CRM will receive the signals directly from all the RMs and will process the input data to locate the TMs and send this location data to CC. Every RM has a memory unit which stores the data received by the CRM. At each appropriate interval, this data with the CRM, as transmitted to the CC, is stored at each RM. The reason for such a step is that as and when the CRM is destroyed in an avalanche, it becomes possible to recover the previously known data from any of the RMs and the CC. This information can be verified for similarity and further authentication so that the consequent dispatching of the rescue team happens at the precise location. The information received at the RMs and CC needs to get compared for the last few time instants for higher precision. This authentication between RM and CC also takes place at non-avalanche conditions for continuous validation purposes. Figure 3.5 describes the flow diagram of the Central Command (CC) unit.

The communication between CRM and CC will happen via satellite and between RM and CRM via line of sight. All the subsystems, TM, RM, CRM, MTM, MCRM, and CC, are synchronized via a satellite clock for precise measurements. MRM and MCRM are the mobile versions of RM and CRM, respectively, which will be used

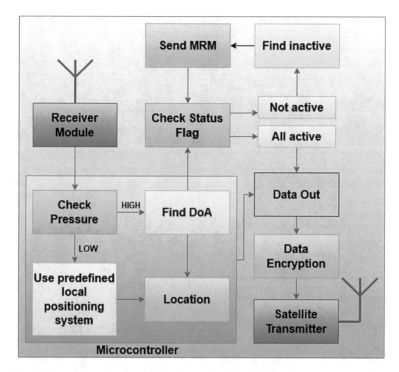

Fig. 3.4 Block diagram of the Central Receiving Module (CRM) unit

when the RM and CRM are malfunctioning due to avalanche or any other cause. The information at the RMs and CC will also be compared with the information available with the MRM and MCRM for the last few time instants for more precision.

3.2 Detailed Description of the System

As shown in Fig. 3.1, the proposed system has one CC unit for the entire system, and separate RM units together cover the whole geographic area of interest. Different geographic areas will be connected to the CC each having its Central Receiving units. In the proposed system, all the TMs are connected to one or the other RM which sends all the information to the CRM.

The TM consists of a microcontroller, a positioning device, a satellite clock, and a pressure sensor. The microcontroller captures the pressure value on the TM and position data from the positioning device. The pressure data will provide information about the present situation of the user; if the pressure is high then the user is trapped under the snow, and if the pressure is normal atmospheric pressure this means that the user is safe. This information is used at the CC to generate a status flag for

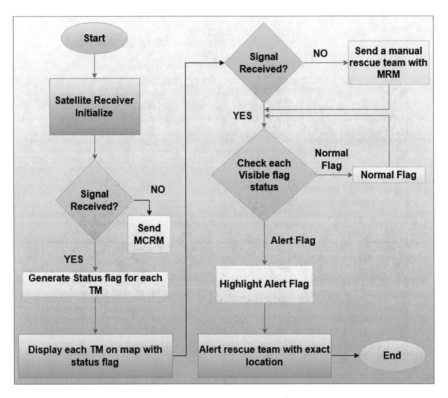

Fig. 3.5 Flow diagram of the Central Command (CC) unit

each TM, and the flag information will determine whether the user is in danger or not. The user equipped with TM may get buried too deep into the snow. For such situations, there is a power amplifier bank which will add more power amplifiers to the transmission network as the pressure increases on the sensor. In this way, the battery life is optimized for the TM while also ensuring that sufficient power amplification is available. Pressure data, specific location data, and the time instance of transmission from the satellite clock are transmitted at a predefined frequency to RM. The specific frequency will be used so that the attenuation due to snow becomes less while keeping the antenna size within the limit.

The RM receives this signal and feeds it to the microcontroller for further processing. The RM units also have a satellite clock, a memory unit, and a status flag. The status flag will indicate whether that particular unit is working well or not. The function of the Receiving Module includes filtering the received signal, adding new data at time t_2 when the signal gets received from TM, and providing status flags to the signal. The duration in which the data transfer happens includes t_1, t_2, pressure data, location (positioning device), and status flag. The memory unit stores the data received by the CRM. There are several RMs installed in a region, and all of these are the recipients of data of different TMs. The data from all these RMs get transferred to a CRM so that processing happens at one centre and also satellite communication can happen from that one place itself. This communication between RM and CRM is of the line-of-sight type and typically takes place at High Frequency (HF) or Very High Frequency (VHF).

At CRM, a check will be performed on the status flag from each RM, such that whenever any RM fails, a substitute in the form of a Mobile Receiver Module can be sent to that location. This won't allow any TM to go off-grid. If any particular TM is buried under the snow, the specific location won't be available as the positioning device works at microwave range and there is a huge attenuation for microwave range inside the snow. So the location of TM will be identified using Direction of Arrival (DoA) data. Otherwise, the location from the positioning device will be used. The output of CRM contains the location data and pressure data which will be stored and encrypted for security purposes. This encrypted signal will be transmitted to a satellite so that it can be sent to the CC. The processed information in CRM is also sent to the RMs.

At the CC first of all the receiver is initialized to receive signals. If no signal is received then it means that the CRM is malfunctioning or is hit by an avalanche, and so a Mobile Central Receiving Module (MCRM) will be sent to that location. Otherwise, if signals are received properly, it generates the status flag for each TM. This generated flag along with the location of each TM will be pointed on the map and displayed on a screen. It may happen that for some of the TMs, signals will not be received by RMs and so they won't be reflected on the map at CC. In such a case a rescue team with an MRM unit will be sent to locate the missing TM. The rescue team will be notified of the location of the TM units whose status flags are in "alert" mode. To cover large areas, such clusters of small areas are clubbed together, and in such a case the CC will be equipped with many different screens such that a very large area gets covered for locating human beings in case of an avalanche. A small prototype consisting of basic components like Battery, Oscillator, Power Amplifier, Transmitting and Receiving Modules are built in the laboratory. The oscillator, as shown in Fig. 3.6, will produce 2 MHz, which will be a power amplifier input.

Through the transmitting module, the amplified 2 MHz signal gets sent to the receiver. A 2 MHz Class E power amplifier also amplifies the oscillator's output signal, as shown in Fig. 3.7. A microstrip antenna helps increase the transmitter's range, and the spectrum analyzer is used as a receiver. The main objective is to design a cost-effective transmitter of 2 MHz that detects trapped humans under the snow with sufficient capacity to connect with RM. The antenna is effective when tuned at the same frequency. Otherwise, the transmission is impaired [9]. A spectrum analyzer measures the power of the signals, frequency, bandwidth, and power spectral density. The Class E power amplifier shown in Fig. 3.8 comprises a transistor, load resistance R-L which acts as a resonator circuit L-C, a parallel-connected capacitance C_p, RF choke, and control.

In the first half period, the switch is on, whereas the switch is off in the other half. The voltage across the switch is zero when the switch is on, and the current is zero with the switch off [10]. The values of the resonator circuit element [11] are

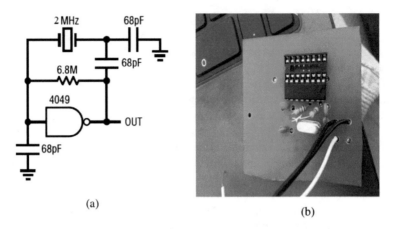

(a)

(b)

Fig. 3.6 Oscillator design **a** schematic, and **b** prototype of an oscillator

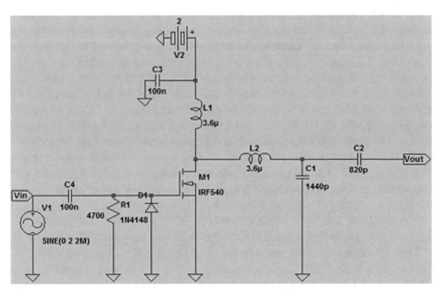

Fig. 3.7 Circuit diagram of Class E P.A.

$$R_L = \frac{8V_{DD}^2}{P_L(\pi^2 + 4)}, \quad L_2 = \frac{8QV_{DD}^2}{\omega P_L(\pi^2 + 4)},$$

$$C_2 = \frac{P_L(\pi^2 + 4)}{8\omega QV_{DD}^2}, \quad C_1 = \frac{P_L}{\pi\omega V_{DD}^2}, \tag{3.1}$$

where $V_{DD} = 2V$ is the drain supply voltage, $P_L = 10W$ is the load power, $\omega = 314$ rad/s is the operating frequency, $Q = 6.08$ is the quality factor of L_2,

Fig. 3.8 Hardware implementation of a Class E P.A.

and $\omega_0 = 12.56$ rad/s is the resonant frequency. From the above values, we find $R_L = 7.5\Omega$, $L_2 = 3.6H$, $C_2 = 820F$, and $C_1 = 1440F$.

To design this antenna, we have printed the design of the antenna on the PCB. The resonant frequency is contingent on the resonant distance and is about $\frac{\lambda_r}{2}$ for a rectangular microstrip antenna, where λ_r represents the wavelength in PCB material [9]. The resonant length is

$$L = 0.49\lambda_r = \frac{0.49\lambda_0}{\sqrt[2]{\epsilon_r}}, \quad \lambda_0 = \frac{c}{f}, \tag{3.2}$$

where λ_r = PCB wavelength, λ_0 = free space wavelength, ϵ_r = dielectric constant of the Fr4 material = 4.7, $c = 3 \times 10^8 \frac{m}{s}$ is the speed of light, and f = transmitting frequency = 2 MHz.

Hence λ_0 becomes 150 m, and resonant length becomes 33.9 m. It is too high. As we want to design the antenna for a range of a few meters, we decrease the antenna's physical length. Here the resonant length of the antenna is 1.62 m. Figure 3.9 has the combined setup of a Class E power amplifier and microstrip antenna. The microstrip antennas are planar and so take limited space in the final setup. The complete TM module hardware is shown in Fig. 3.9.

3.3 Results and Discussions

In this module, we generated 6 V, 2 MHz sinusoidal signal, and the output waveform is shown in Fig. 3.10. This signal will be passed on to a Class E Power Amplifier as input 1.83 V, 2 MHz sinusoidal signals, and a resulting output 3.9 V, 2 MHz. Thus, the voltage gain becomes 2.13 or 6.57 dB. The FFT plot of a Class E power amplifier is shown in Fig. 3.11; we can see the peak at 2 MHz frequency, which shows that the output is generated at 2 MHz frequency. The signals are also clear even when TM

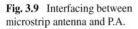

Fig. 3.9 Interfacing between microstrip antenna and P.A.

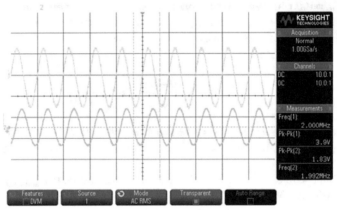

Fig. 3.10 Waveform of Class E P.A.

is put inside the half feet of snow, and can detect up to a distance of 5 m above the snow surface.

The complete prototype is shown in Fig. 3.12. The Spectrum Analyzer receives the transmitted signals. Figure 3.13 shows the output of the spectrum analyzer. Setting the start and stop frequencies at 1 MHz and 3 MHz, respectively, we find the peak at 2 MHz that conveys antenna transmission at 2 MHz. We have obtained the power spectral density of −115 : 66 dBm/Hz at the 2 MHz frequency transmission. The Class E power amplifier shows a resulting gain of 6.57 dB.

This chapter proposed system architecture to identify and locate humans trapped in the snow using a pre-formulated local positioning system. A small prototype was also prepared and tested in the lab using a cost-effective 2 MHz snow beacon. The energy-efficient transmitter modules transmit signals when the pressure sensor measures the stress from the outer environment. The transmission pulses are generated with the

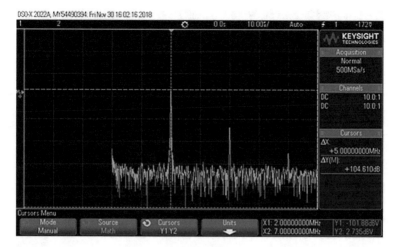

Fig. 3.11 Class E P.A. FFT signal analysis

Fig. 3.12 System setup

help of an oscillator circuit. Signals are amplified using the Class E amplifier. The signals are transmitted with the help of a designed patched antenna. The received signal power and power spectral density are analyzed with the help of a spectrum analyzer. The practical implementation of an RF-based tracking system is discussed in this chapter and Chap. 2. A computer vision-based object tracking recognition technique to monitor an area is discussed in Chap. 4. Detection algorithms are applied to real-time captured frames by onboard processing embedded units. This study aims to recognize the individual identity and notify the authority of the security breach.

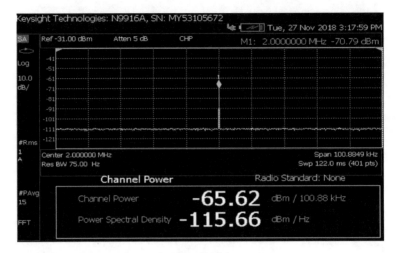

Fig. 3.13 Transmitted signal detection using spectrum analyzer

The facial and object detection-recognition method and providing controlled access using the smart door closer system are discussed in Chap. 4.

References

1. ASTMF1491-93: Specification for an avalanche beacon frequency (2002)
2. Upadhyay, J., Rawat, A., Deb, D., Muresan, V., Unguresan, M.L.: An RSSI-based localization, path planning and computer vision-based decision making robotic system. Electronics **9**(8), 1326 (2020)
3. Deb, D., Rawat, A., Khanna, N., Agrawal, S., Radadiya, H.: Smart imperceptible door closing systems with integrated monitoring (June, 30 2017), Indian Patent 201721021291 A
4. Prajapati, U., Rawat, A., Deb, D.: Integrated peripheral security system for different areas based on exchange of specific data rates. Wirel. Pers. Commun. **111**(3), 1355–1366 (2019)
5. Prajapati, U., Reddy, B.S.K., Rawat, A.: Modeling a data rate-based peripheral security system in GNU radio. In: Integrated Intelligent Computing, Communication and Security, pp. 353–361. Springer, Singapore, Sep (2018)
6. Prajapati, U., Rawat, A., Deb, D.: A novel approach towards a low cost peripheral security system based on specific data rates. Wirel. Pers. Commun. **99**(4), 1625–1637 (2018)
7. Upadhyay, J., Deb, D., Rawat, A.: Design of smart door closer system with image classification over WLAN. Wirel. Pers. Commun. (2019)
8. Surve, J., Mehta, V., Rawat, A., Kamaliya, K., Deb, D.: Low-cost 2 MHz transmitter for the detection of human trapped under the snow. In: Lecture Notes in Electrical Engineering, pp. 41–50. Springer, Singapore (2019)
9. Orban, D., Moernaut, G.: Orban microwave products. The Basics of Path Antennas, XP-002764118, RF Globalnet (2009)
10. Eroglu, A.: Introduction to RF Power Amplifier Design and Simulation. CRC Press (2018)
11. Yousefi, M., Kozehkanani, Z.D., Sobhi, J., Azizkandi, N.N.: Improved efficiency 2.4 GHz class-e power amplifier with improved controlled output power. Indian J. Sci. Technol. **8**(23) (2015)

Chapter 4
WLAN-Based Smart Door Closer Design with Image Classification

This chapter discusses an innovative mechanism for object and person recognition and detection, which utilizes a robotic control system to restrict and monitor sensitive or restricted areas from events such as intrusion. This chapter aims to initiate the surveillance or monitoring right from the first point of interaction and verification. The system is designed to perform a multitude of actions which include the following: (i) to register the scenario, (ii) understand the environment, and (iii) provide the control access signal to all other peripherals such as magnetic door closer unit or other robotic systems. The innovation also ensures that the overall installed system is imperceptible, smaller in size, and remotely operated with low power consumption. The algorithm enhances security in the monitoring area by deploying a robotic agent with image recognition and analysis capabilities.

A dual-purpose system is presented in this chapter, which serves as a door closer and equally for surveillance purposes. The currently deployed surveillance systems store a large amount of data, thereby consuming large memory spaces. This chapter's novel feature is that object identification and classification are performed for the desired area, and the provision for controlled access. This system uses a neural network-based learning algorithm before providing any instruction to the hardware. This novel design on the door closer system and the robotic system with image processing provide enhanced security advancement. This novel design holds embedded devices within a hydraulic door closer, which is almost similar in size to a conventional door closer. It isn't easy to recognize it from a distance to be any different from a regular mechanical door stopper. To provide enhanced security, the results are stored on a remote server through the wireless network.

Section 4.1 discusses the present scenario of security systems and problems. Section 4.2 describes a facial detection and recognition algorithm implemented on low-power devices. Facial detection techniques depend on various parameters. The enhanced identification technique of individual detection is described in Sect. 4.3.

Section 4.4 is focused on the hardware implementation, and Sect. 4.5 explains the system flow and module placement inside the door closer system. Data acquisition and analysis over the wireless local area network is described in Sects. 4.6 and 4.7.

4.1 Background and Motivation

The present growth of the economy in urban areas warrants higher and efficient security. Ensuring an individual's privacy and security at the same time is a challenging area of work. Surveillance and security systems play a significant role in reducing crime rates by collecting strong evidence and creating an impression in the intruder's mind of being observed. Studies have shown that the areas with higher crime rates typically cannot be developed in terms of economy [1] or see significant stagnation in development. The crime patterns are also changing with time and with an increasing rate of crimes [2]. Such analysis of crime patterns, detection, and predictions are actively studied by researchers for many years [3–7]. Such studies reveal strong anecdotal evidence of surveillance cameras facilitating offenders' detection and conviction and curbing crimes.

An onboard embedded device can be a useful tool to analyze the data coming from the camera or the sensor unit. Several open-source low-cost, powerful, and energy-efficient embedded devices are available to process the data. This system supports different library functions that analyze raw data coming right from the sensors. The integration of open-source software packages with open-source hardware provides a re-configurable and licensed free system solutions. The Raspberry Pi is one such open-source development board. The selection of a development board for any application is crucial because the software package libraries are chosen as per the chipset specification. And the software architecture with hardware integration plays a significant role in application accuracy and performance. This is the first step in the selection of the core aspects of processing before finalizing any design.

The selection of the sensor is another significant part of this study. All the theories and analysis are applied to the data collected from the peripheral sensor unit. There are a wide range of embedded modules in the hardware development environment with the capability to handle medium to moderate computational processes in a limited space, but with all necessary add-ons. This small size of hardware should have the capability to perform the application of deep learning, computer vision, and Artificial Intelligence (AI) programs, and operate in the Linux environment. Also, the embedded device needs the capability to operate from a remote computer, so that the system can be implemented in critical areas and also be operated remotely, providing added design flexibility.

The hardware specifications and the dimensions, chipsets (electronic components in an integrated circuit for management of data flow), and power consumption level vary as per the model. The Raspberry Pi Zero W is the smallest among all available hardware in this family and is a possible fit for the constrained environment envisaged in this work. The power consumption levels vary from 1.3 to 1.5 W. This system is also compatible with a portable and lightweight Pi camera which captures frames with wireless area network access connectivity and a serial interface protocol for usage in image processing, machine learning, and surveillance.

A low-cost biometric sensor may be interfaced with Raspberry Pi devices so that a person's encoded biometric characteristics are stored in the cloud storage (Azure cloud) [8]. Data privacy is an essential factor in the increasingly digital world. The footage received from a close circuit television camera (CCTV) at the crime scene can help to investigate crime cases [9, 10]. Installation of intelligent surveillance systems has seen a rapid spurt owing to video analytics, and advanced data mining techniques [11, 12]. The government or the private sector uses such systems, but one may use a pocket-friendly system for personal security purposes. The efficiencies of such devices vary and are mostly dependent on how the system is set up and how it records data [13].

Many systems are directly connected to the wireless nodes, and a secured connection prevents alteration of any sensory hardware like CCTV, with the movement sensed by measuring the change in data rate [14, 15]. The shift in data rate is measured and implemented. The software implementation is simulated with open-source software-defined radio software [16]. A low-cost frequency beacon identifies the target's location inside a dense environment [17]. This type of system implemented in remote areas senses the person's movement in sensitive regions. A GPS signal gets multi-path reflection from the different objects inside the indoor environment. Getting an accurate position of any target of the robotic agent is a difficult task. Upadhyay et al. proposed an RSSI and computer vision-based indoor navigation techniques to localize an object in a closed environment. The RSSI signals captured at the receiver side are classified into the mapped position [18]. For an accurate result in a dynamic environment, a neural network-based classification approach gives promising results to localize a robotic agent.

One of the major issues in computer vision is pedestrian detection, which also affects lifestyle. Dollar et al. discuss pedestrian detection dataset and studies the statistical properties in terms of occlusion, position, and size. The performance of prime detection methods for six different datasets is presented [19]. Loukhaoukha et al. proposed image encryption techniques in which the logical operation XOR between even-odd row and column of image pixel RGB values, is performed. Besides, image vector transpose is applied between the even rows and columns. This encryption technique is faster compared to other methods and can be implemented over an embedded device having low computational frequency [20]. Upadhyay et al. described the stealth monitoring system's implementation over a compact embedded system [21]. In the proposed method, objects like persons, animals, tablets, laptops, chairs, etc., are recorded in real-time video stream data, captured through a hidden camera on the door (with a door stopper mounted on it). The analysis involves image classifiers: (i) object detection classifiers and (ii) face detection and recognition classifiers. These two classifiers work independently of each other and feed intelligence to the server for further processing.

4.2   Face Recognition and Detection

When motion is detected outside the room, the module meant for object detection and classification attempts to identify the object, and after that, the second classifiers are applied. The module meant for facial detection and recognition is used to identify the specific person, and based on that the door is opened or stays closed as per policy. In this system, the identification is based on facial recognition and the surveillance system only records necessary data of person and events, rather than uninterrupted video files, so as to save battery. The facial detection and recognition algorithm is shown in Fig. 4.1. This flow graph describes the working of the facial detection algorithm.

The images get captured from the live frame collected from the Raspberry Pi camera module or CCTV unit. An image database with folders specific to individual persons containing respective images useful in training the neural network is essential. The first step is to be able to segregate and recognize the face in the image pixels itself, and then convert it to the specific size so that the background envi-

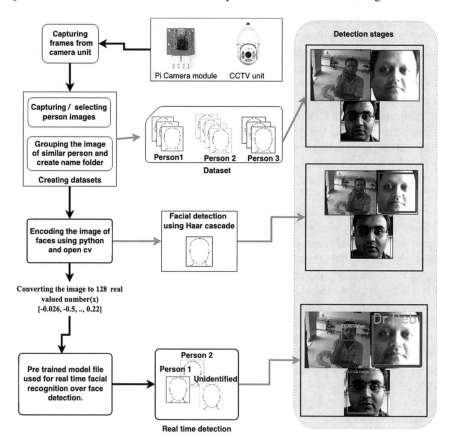

Fig. 4.1 Stages of facial detection and recognition method

ronment does not affect the appropriate inference or the result. Then facial images are encoded with the help of the OpenCV-Python library function. The OpenCV is an open-source library function for digital image processing which enables an image/frame to call upon the Python coding framework for further processing. After detecting the facial region, the program converts the images with the real-valued vector for each face used as a reference. The algorithm conveniently labels the identified person as their database folder name. In case of a mismatch, the inference is that it is an unauthorized person. The haar frontal face architecture used in this algorithm is faster than convolutional neural networks. This architecture mainly detects the facial area from an image. The green area in Fig. 4.1 shows the detection stages: first extract the image from the frame, second identify the face area, and third detect individual using a pre-trained model.

4.3 Individual Identity Verification Using Key-Point Matching Algorithm Technique

The facial detection algorithm is flawless in adequate lighting conditions. While the same algorithm may not promise an accurate result if the facial area is covered by a mask, it is difficult to control restricted areas if individuals wear a mask. The key-point matching technique compares similarity points from the real-time image and database image. Before applying the algorithm, the object detection algorithm detects the person area and extracts it from the image; this image will be compared with the database image. The system flow for the detection technique is shown in Algorithm 1. Individual images are collected and labeled in the database. Images are collected at 0.12 s per sample. Multiple images are collected to minimize the lighting condition effect in sampled data. The object detection algorithm identifies the number of persons in the image. The detected person area is collected and stored differently in the image. These images are further processed by a facial detection algorithm. The facial area is extracted from the image for further processing. The facial area will be detected if the key feature point areas such as eyes, nose, lip, and jaw dimension area are clearly visible at the time of image capturing. Extracted images are classified with the help of a pre-trained model. As the object detection algorithm identifies a person even if the person wears a mask, the number of detected persons by the object detection classifier and the facial detection classifier count should be the same. If the facial area is not detected by the classifier, then this count will not match. In this case, the point matching algorithm identifies individual.

The key-point matching algorithm extracts the individual person image area and stores it on local memory labeled with "Person Number". These images become templates to identify their matches in the image database. An image histogram is generated to identify the color distribution in images. A low-resolution image has higher gray scale distribution. For accurate detection, captured images have higher resolution. The camera module is configured to get higher resolution images and transfers these images through a wireless and local area network to the control center. The Scale-Invariant Feature Transform (SIFT) algorithm can be used to find the features from the images. The feature matching algorithm identifies the similarities in the images such as pattern, color, and orientation. If the features are not matched in all the database images, then the person image is captured again for detection. The distance of valid matching points in the images is set to 800 Euclidean pixel distance and the number of valid matching points should be greater than 300.

Algorithm 1: Point matching based person identification

Data: Individual labeled images with face mask
Result: Image comparison using key-point matching
initialization;
while *Person detected by the object detection classifier* **do**
    instructions;
    if *Facial detection algorithm detects the facial area* **then**
       | Apply facial detection and recognition algorithm;
    else
       Create a database and capture ten images within 2 seconds;
       Store images in database, label it with person name;
       Discard the data at the end of the detection session;
    end
    if *Person is detected by classifier* **then**
       Wait and store the detected person images with their name;
       if *Person count = facial detection count* **then**
          | Store the images with time and accessing area;
          | wait for new image for detection;
       else
          | Apply key-point matching algorithm;
       end
    else
       | Apply key-point matching algorithm over captured frame image;
    end
end
while *Key-point matching algorithm applied* **do**
    instructions;
    if *Capture individual person bounding area from image* **then**
       select real-time image as template image for matching with database;
       Obtain image histogram;
       apply SIFT detector to find key-points;
       match the key-point based on point pixel distance match value;
    else
       | Capture person image again;
    end
    if *Point distance $\geq$ 800 and matching* **then**
       Result is valid image and Image matches with the database;
       Show label of person image and store it with time stamp with access area detail;
    else
       Person count $\neq$ detection count or No match in database;
       Send notification to authority;
    end
end

These threshold points are tested for different lighting conditions and both camera modules, i.e., door closer unit camera and CCTV camera modules. After the detection, the person count is again matched with the person detected by the object detector classifier. The moment at which the total number of facial and key-point matching detector algorithm equates the individual count, the overall process of the detection algorithm is considered as completed. If the count value does not match, then the undetected images are sent to the control center to notify the person unidentified in the region where the images are taken.

4.4 Smart Door Closer System

A hydraulic door closer system assembled in a small region, and configured with an integrated embedded system, is described herewith. The proposed development has two main components: (i) a hydraulic door closer system with an embedded system installed at the top right part (but not visible) and (ii) a magnetic door lock at the door's top left corner to control the door closing and opening, as shown in Fig. 4.2.

The magnetic lock generating a force of 1200 lb is controlled by a 12 V DC relay control unit. A control pulse given to the relay cuts off the magnetic field and opens the door. The embedded system connects to a local Linux operating system-based server over Wireless Local Area Network (WLAN). Raspberry Pi zero is used since its limited size is ideal. However, with a restricted processing power of the Raspberry Pi zero units, only a limited amount of image classification is possible. The device only transmits live frames when it detects any motion in the camera aperture area. The facial recognition algorithm takes the frame-by-frame images from the video stream to analyze and decipher individuals based on their pre-trained model. As shown in the functional diagram in Fig. 4.3, the system is integrated with a single camera module. The 230 v–50 Hz A.C. supply is rectified to a 15 V D.C. voltage level. Further, this 15 V supply is regulated by 5 and 12 V voltage regulator units to power battery (4000 mAh) and relay unit, respectively.

The battery unit provides a constant supply to the embedded module. If the primary source is off, then the battery unit will ensure regulated power for up to 10 hours. The magnetic door lock is controlled by the controlling pulse of the embedded module through the 5 V D.C. relay unit. After applying a controlling pulse, the relay switches to a normally open condition. The embedded unit connects to the database server through a wireless data link.

Further, a remote operating system collects the information through the database server and processes such information if necessary. In the case of motion detection by the embedded unit, the event is captured and analyzed. When the authentication control pulses are provided to the relay unit to open the door, the system only collects the individual facial images and time of entry to limit memory utilization. This processed data will be stored at regular time intervals to maintain the performance of the embedded unit that may be affected by the overflow of cache memory. The cache memory is limited for any embedded device, and stores frequently used instruction location and memory map. In case of process overload at the onboard embedded

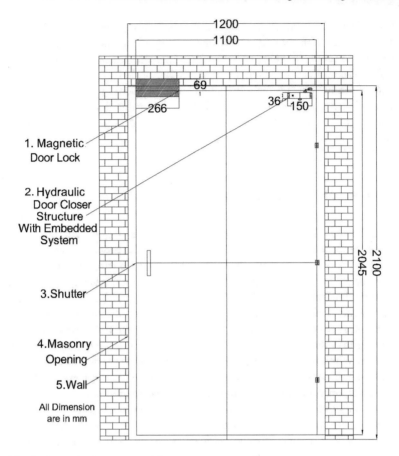

Fig. 4.2 Hardware placement to control area access

unit, the system starts to stream frames to the server, and the control center processes the data and sends the control signal via the database server.

Figure 4.4 describes a detailed sectional drawing of the patented design. The components installed in the door closer while manufacturing the system are (i) Microcontroller, (ii) Camera module, (iii) Communication module, (iv) Microphone module, and (v) Magnetic Hall effect sensor. The positioning of the electronic components ensures that the actual dimension of a hydraulic door closer system remains identical to the conventional door closer module. The Hall effect sensor measures the displacement of the piston during movement. If the sensor measures the displacement during the closed position of the door, the embedded system captures the frame and notifies the authority about a security breach. The microphone is at the top of the body, so the vents are not visible to anyone.

An extended view of the door closure system is shown in Fig. 4.5 describing the actual placement of the Raspberry Pi zero unit used as an embedded device with the Pi cam interface for the visual data. As seen in the diagram, the embedded unit is fitted inside the door closer unit.

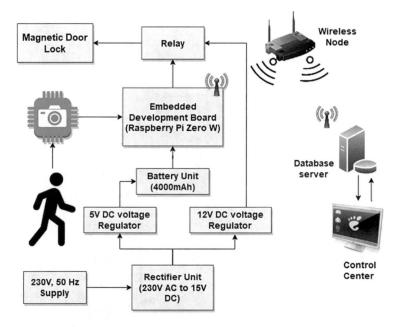

Fig. 4.3 Hardware functioning represented as a block diagram

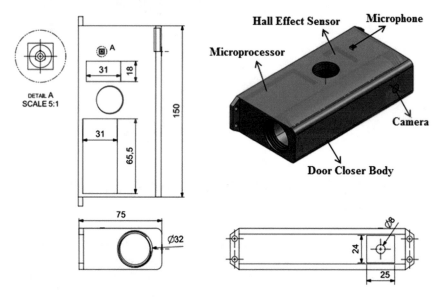

Fig. 4.4 Placement of hardware inside a uni-body door closer case [22]

Fig. 4.5 Placement of embedded unit and camera module in the door closer unit

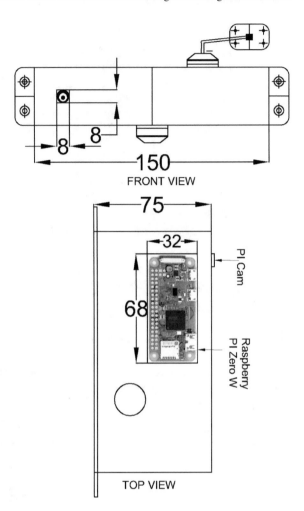

FRONT VIEW

TOP VIEW

4.5 Working Principle of the Smart Closer System and Robotic Platform

In the proposed module as the embedded system boots up with a predefined code for motion detection, wireless data transmission, and facial recognition with magnetic sensor control based upon classification techniques and sensitivity, the different steps involved in the signal flow are shown in Fig. 4.6. First, the system detects the motion in the surrounding area using the camera mounted on the top right side of the door. The object movement in the pixel area performs the detection. The camera starts recording the images frame by frame and generates a video file which is then called into the code to analyze the frame-by-frame images, and then the face detection algorithm is applied to those images to extract the face and compare it with the facial image stored in the neural network training model.

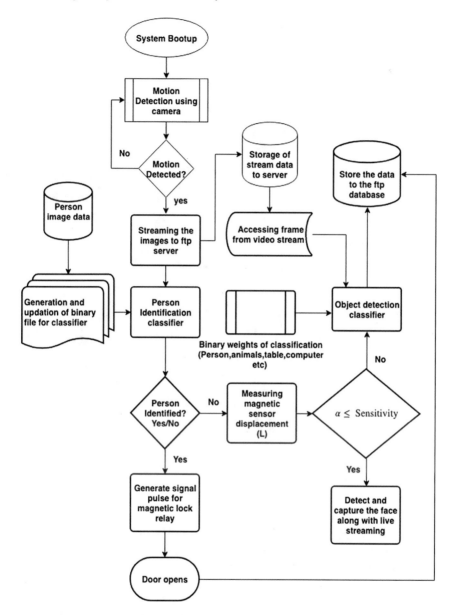

Fig. 4.6 System flow graph for area monitoring

Fig. 4.7 Actual dimensions and placement position of the surveillance component

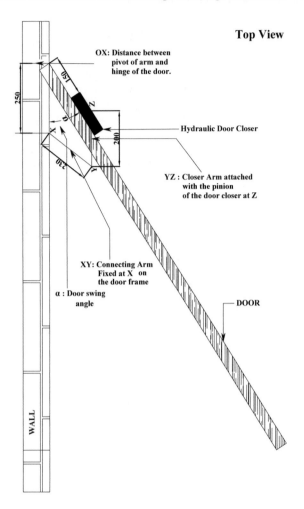

If the face matches with that in the training model, the relay of the magnetic lock opens the door. The dimensions and positioning of components required for analysis are given in Fig. 4.7. If the person is not authorized or unidentified, the system performs surveillance and records the area.

On identifying any significant movement by an unauthorized person or object, the door senses the movement through a magnetic Hall effect sensor (as already described in Fig. 4.4). The piston's displacement in the chamber area of the hydraulic door closer system ensures that the inner side of the camera also starts recording the footage. Initially, the teeth of the pinion in contact with the teeth of the rack get relocated. The teeth not connected with the rack then shift toward the position of being in proximity to it. The rotation of the pinion makes the piston move inside the chamber. The magnetic sensor module fixed in the door closer setup must be at a location that would receive a magnetic field whenever the door opens. An empirical relationship between the pinion rotation α and the linear movement of the rack is

$$L = \frac{\alpha \pi D}{360°},$$

where L is the length of the arc of contact or linear displacement of the rack measured by the magnetic Hall effect sensor, D is the pinion's pitch circle diameter, and πD is considered as the circumference of the pinion. If $\alpha \leq sensitivity$, the classifier only monitors the area, and if this threshold is violated, then streaming gets stored into the database.

The proposed system, an outcome of a patent disclosure, also works as security surveillance. However, in the present case, motion detection in the camera aperture area initiates the system to monitor and recognize the person's face. The appropriate video frames turned on on demand are transmitted to the remote Linux-based server where the object detection and recognition classification happens over real-time frames. This classification identifies the object, then recognizes it from the pre-trained model. The trained model classifies the objects under observation like a person, chair, laptop, cell phone, animal, car, monitor, etc. The surveillance system largely depends on the design of the door closer system. For desired results, the placement of the component should be precise. Because of the smaller size and constrained region of the design, the exact location and orientation of the embedded systems, Raspberry Pi Zero W, Pi camera module, and battery unit positioned along the other side of the barrel containing the pinion, are critical.

The camera position ensures coverage of a maximum aperture area. Such a smart system gets incorporated universally with a traditional door closer without any significant structural modifications. The proposed smart imperceptible door stopper design contains an inventive step wherein the electronic components come to life at specific well-programmed angles of the door opening, which corresponds to the position of the liquid inside the door stopper. The synchronized functioning ensures interdependence of the two subsystems: (i) the traditional door stopper and (ii) the embedded systems incorporated into the door stopper.

Storing the image and video in a local server without any data protection techniques can lead to data security threats. The process images are important assets for many applications when this algorithm is deployed inside an educational or financial institution. The security algorithm needs to be modified on a day-to-day basis, and the passing of an encryption-decryption key over a wireless medium is a crucial task. In this proposed method, the data and the keys are both encrypted before transmission. The results of image encryption-decryption are shown in Fig. 4.8.

The embedded units in the door closer unit and the surveillance robotic platform encrypt the data before transmission over a remote server through a wireless medium. The encryption key is transmitted on a different channel and is only transmitted after the authority sends a specific request to the monitoring hardware. Before transmitting the encryption key to the control center, the algorithm converts the key into a single-time downloadable encryption file through the server link. The authority first downloads the encryption file and then performs the decryption process.

Fig. 4.8 Image
encryption/decryption before
transmission

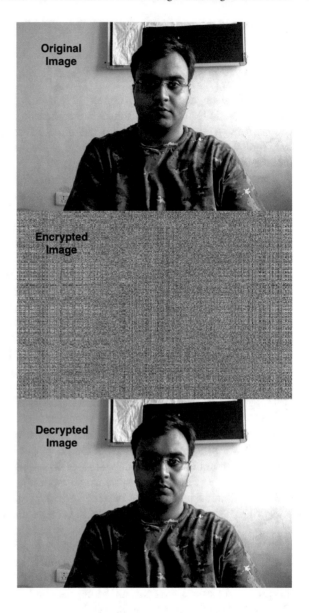

The robotic agent analyzes area and critical angle information from the places that are not easily detected by the door closer units. The agent analyzes the image captured from camera unit and analyze the individual their surrounding by applying the facial/object detection and recognition algorithm. Several techniques navigate a robotic platform inside an indoor environment, such as a 3D vision sensor or Light Detection and Ranging (LIDAR)-based simultaneous localization and mapping. However, the sensors have their own limitation for low-power robotic platforms.

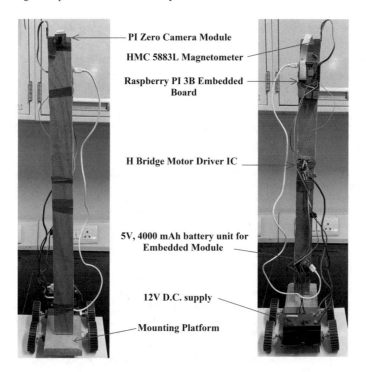

Fig. 4.9 Robotic platform for surveillance

The Received Signal Strength Index (RSSI)-based indoor navigation techniques are used to navigate the robot inside the indoor environment. The navigation technique is explained in detail in Chap. 5.

The robotic platform for surveillance is shown in Fig. 4.9. The embedded hardware module Raspberry Pi 3B is used as an algorithm processing device. The module is interfaced with the HMC 588L magnetometer sensor to measure angular displacement concerning the true north position. A Raspberry Pi camera is interfaced with a processing unit to capture high-resolution live frames. The optimal path planning algorithm is used to navigate a robot between the current position and target location. The path following algorithm manages the motor kinematics through the H-bridge motor driver integrated circuit (I.C.). The 12 V battery unit provides a constant power supply to the motors, and a separate (5 V, 4000 mAh) rechargeable power cell unit powers the embedded module.

Algorithm 2 systematically describes the functioning of the robotic platform. The agent analyzes the events through the camera module and applies the detection algorithm while following the path from source to destination, decided by the path planning algorithm. The images are processed, and the results are stored on board. The agent's objective is to identify and analyze each captured frame to detect sensitive objects like a weapon or firearm and the person carrying them. In the case of detection, the agent transfers the robot's handling to the authority and broadcasts the live frames with it. The robot tries to follow the object/person by keeping it at the center of the image. The robot decides to move right or left by tracking the object in an image and calculating the image center's Euclidean distance. The result of processed images is stored on board and on the remote server after encryption.

Algorithm 2: Decision-making movement algorithm for robot

Data: Captured images from live frames
Result: Sensitive object and facial detection-recognition
initialization;
while *Person or object detected by the object detection classifier* **do**
  instructions;
  if *Facial detection algorithm detects the facial area* **then**
    Apply facial detection and recognition algorithm;
    Store the images if the person is unknown;
  else
    Identify other objects;
  end
  if *Suspicious activity detected by the robotic agent* **then**
    Wait and store the detected person images and objects;
    if *Weapon detected in the monitoring area* **then**
      Broadcast the live frame to the control center;
      if *Detected sensitive object/person is moving* **then**
        Track the object/person inside the image;
        calculate the Euclidean pixel distance concerning target center and image center;
      else
        Send the live frames to authority;
      end
    else
      Move to the next target;
    end
  else
    Move to the next target location decided by the optimal path planning algorithm;
  end
end

4.6   Data Acquisition (DAQ) of Smart Door Closer over WLAN

When in need of higher computational capability, the embedded system works as a video data transmitter, and the server acts as a processing device. In such a situation, the analysis of the live data frames and the processing of the object and facial detection operations take place at the server-side, and the door locking embedded system receives a signal to open and close the door from the server-side. The Raspberry Pi embedded module is the onboard processing unit placed in the monitoring area.

This system cannot transfer the frames over the remote server without configuring the streaming and wireless modes. The program at the remote server collects the frames stored in the local data memory. Frames, after all, are a collection of images, and so if data transfer happens without configuring the wireless protocol, there could be a loss of critical images during the transmission. The motion configuration is the package library for video transmission installed in an embedded module. For data retrieval from a Raspberry Pi zero unit functioning as a remote wireless camera, there is a need to have a motion configuration library with parameters which are the set-points for image transmission. Continuous streaming with frame transmission cannot happen without setting this individual value. The first parameter is the daemon that releases the system terminal after the motion detection by the default algorithm of the library when the image sensitivity changes by moving the object. The algorithm senses the movement and sends a command to start recording the frames on the local memory space. Parameter setup_mode limits webcam/Pi cam during direct transmission. Image quality is critical for the detection algorithm, and the image height/width value entered for each frame ensures a negligible delay between frame capture and the record on the system/remote server. With the transmission image quality set at 70 in this proposed method, the detection algorithm identifies the object with the range of 30–70 picture quality.

Pre_capture and post_capture variables provide the differential number of the images before recording the actual frame. From practical experience, it suffices if the system captures at least five frames after the motion algorithm stops sensing any movement in the deployed region. Captured_frame rate refers to the frames per second stored in the system. The picture storage happens in either .jpeg or .png file format, and the Frames in different formats. Similarly, the frame rate is set by stream_maxrate. The frame transmission happens over the network, and the remote server collects the frames at the embedded unit IP address: port address. This port address can be 8080, 8081, or 8082. For security purposes, the streaming is limited by access control using stream_authentication, and the streaming data is stored remotely. The live video stream generated from an embedded device is available through the IP address (10.10.64.179:8081), as shown in Fig. 4.10. The IP address is the same IP address of the embedded unit connected to the WLAN, and the 8081 is the port address on which the system broadcasts the frame over the network. The capturing of the frames involves an authorized person, and the embedded unit ensures secure transmission, preceded by frames and image encoding. Figure 4.11 represents the live streaming, and the right side of the area shows the object detection on the frame.

The IP shows a frame rate gap of two frames per second to process a large number of frames through the classifier. This delay reduces the choice of a smaller frame size for the video while processing. As an outcome, the system provides the identified results such as persons and laptops. The results of the facial detection and recognition algorithm are the input for the object detection and recognition algorithm. Figure 4.12 depicts the practical outcomes of facial recognition, and then the object detection classifiers applied over facial recognition. A smart door closer-based monitoring system solves several limitations of a regular security camera. It provides a dual mechanism of automatically closing the door and imperceptibly securing a particular area. The system is enclosed within a compact structure, thereby making it immune to external damages such as contamination and tampering.

Fig. 4.10 Real-time frame
transmission over WLAN

Fig. 4.11 Applying object
detection-recognition
algorithm over a captured
real-time frame

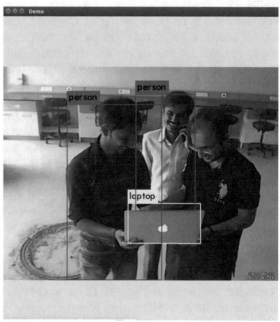

Fig. 4.12 Applying facial detection-recognition algorithm over the result of object detection algorithm

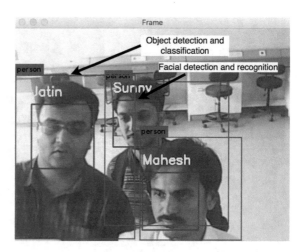

4.7 Analysis of Images Received from CCTV Units

The live frames captured by CCTV units and the door closer unit are transmitted to the control center for image processing and storage purposes. For fast, secure, and efficient real-time data processing, the system first stores the live frames to the remote server through Real Time Streaming Protocol (RTSP) over a local area network. The RTSP protocol can provide the control of the streaming device through the server without transmitting data. After sensing the activity by the motion detection algorithm, the embedded units on the door closer system start processing the images and transfer the frames to the control center for monitoring purposes. The frames collected by the CCTV units are processed at the control center, and object-facial detection and recognition algorithm is applied over that. The remote server analyzes the data by applying suitable classifiers and passes the decision of pulse generation to the client device to decide whether to open the door or not. The proposed system detects and identifies the person with their area accessing time with duration. These data are stored as a comma-separated value file in the local memory. At regular time intervals, these recorded files are stored at the remote server. Before applying the facial detection and recognition classifiers, the system applies object detection and recognition identifier as a surveillance system for storing required events only rather than storing all the events. The images captured from the CCTV unit camera are shown in Figs. 4.13 and 4.14.

The object detection and recognition algorithm is applied over the captured images from two sites, and the detection results are shown in Figs. 4.15 and 4.16.

The individual objects from the image extracted from corresponding images are shown in Figs. 4.17 and 4.18. The facial detection and recognition algorithm applied over the captured images and detected image results are shown in Fig. 4.19. One person without a mask is detected as an "Unknown" person because his image is not present in the database. The algorithm sends the image as an "Unkown" person

Fig. 4.13 CCTV images captured from monitoring area site 1

Fig. 4.14 CCTV images captured from monitoring area site 2

and unauthorized movement in a restricted area. The facial detection algorithm that shows no results for the remaining person image is shown in Figs. 4.15 and 4.16 as the key features of the facial zone like nose, eyes, or jaw area are covered with the mask.

The key-point matching algorithm matches the captured image with the database. One of the results is shown in Fig. 4.20 and 4.21.

The main captured images on the left side in Figs. 4.20 and 4.21 are compared with ones in the database entry of the image shown on the right side. The results

Fig. 4.15 Object detection and recognition classification over CCTV image (Site 1)

Fig. 4.16 Object detection and recognition classification over CCTV image (Site 2)

show no match with the database image in Fig. 4.20 whereas the image matches with a valid database entry as shown in Fig. 4.21.

The label of valid entry of the image is reflected with results; in this case, the image matches with the employee id "Emp 1008 J Upadhyay.png". As shown in Fig. 4.22, the two images are database images which indicate that before deploying the model for detection, the individual image result should not match with each other. Two images of an individual in Fig. 4.22 show no matching points even with the

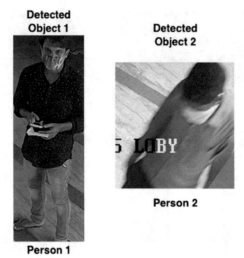

Fig. 4.17 Extracted individual object from Fig. 4.15

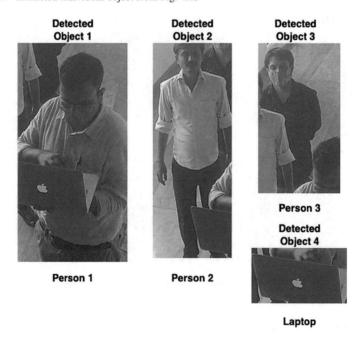

Fig. 4.18 Extracted individual object from Fig. 4.16

Fig. 4.19 Result of facial detection and recognition algorithm for Fig. 4.14

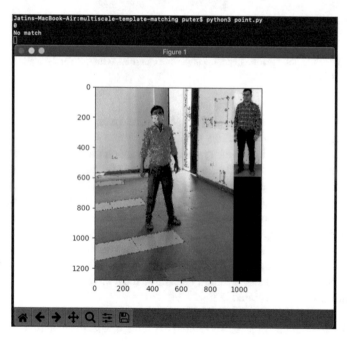

Fig. 4.20 Result of point feature matching algorithm. (Result output: No matching with database and person identity is unknown)

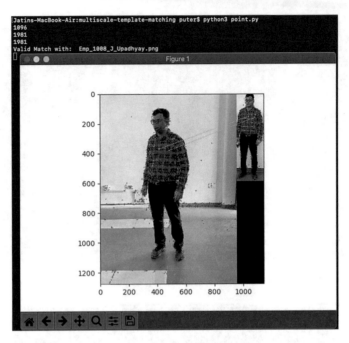

Fig. 4.21 Result of point feature matching algorithm classifier. (Output: Valid match with database and person identified)

same background. The valid match of a person shown on left side in Fig. 4.22 with the database image is shown in Fig. 4.23.

The transmitted data is first encrypted before transmission, and also the transmission link is password protected. The authorized system has to first access the data over a secured link via an authorized login identification and password. After that, the received data is decrypted with the key that is generated at the time of encryption.

This encryption key is sent to a remote server through a separate secure link. The generated file is for single use only so after accessing the file one time, it cannot be downloaded again. To access the encryption key again, the decryption system has to request the control center. For security enhancement, the algorithm is specifically trained for sensitive objects to identify sensitive movements made by any person in the monitoring area illustrated in Fig. 4.24.

The object detection and recognition algorithm is trained to detect the firearms and knife-type weapon objects. As seen in the same image, the detection is stored in the database as comma-separated values. The person's location is identified based on the camera location and the object position inside the image. The proposed system provides solutions to applications that need cognitive monitoring with controlled access to IoT devices. The proposed method identifies a sensitive activity, sensitive objects, person movement, and persons' faces. The detection methods identify the person even if the individual facial area is covered with a mask. Individual access is

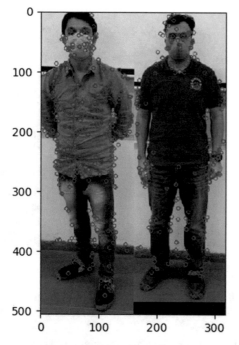

Fig. 4.22 Database image comparison. (Result: Not matched)

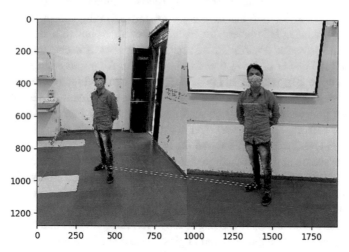

Fig. 4.23 Result of point feature matching algorithm classifier for database image shown in Fig. 4.22. (Output: Valid match with database and person identified)

Fig. 4.24 Sensitive object
detection

limited by the use of advanced door closer devices. The methods show potential to work in requirements like low power, space, and stealth monitoring which meets the requirements of IoT devices.

This chapter highlights an image classification-based door stopper system, a novel solution with a dual mechanism of security systems and cognition-based neural networks. The system is economically efficient owing to its minimal space requirements and can be implemented universally. Facial detection is applied to detect individual identity. The proposed mechanism serves the purpose of efficient storage usage and draws minimal power of 1.3–1.5 W from the 4000 mAh power source. This system will stand up to 10 h without any power backup. The results are stored on a remote server for data security. In the subsequent chapters, we discuss the real-time application as a financial technology to secure an area without human interaction. This will lead to security advancement over conventional systems.

In this chapter, we have discussed the real-time image extraction, image detection, classification and identification techniques as well as remote data transmission over a WLAN for security purpose. The image received over the door closer unit is processed by the onboard embedded device, and the control center processes frames captured by CCTV units. The algorithm uses the pre-existing surveillance system and provides controlled access to the monitoring area using a smart door closer system. Robotic agents analyze the area and angles that are not covered by the door closer system and CCTV units. The object detection and recognition algorithm processes the image and finds the objects inside the images with a pre-trained neural network detection model. The output of the detection algorithm is input for the facial detection algorithm. The facial detection and recognition algorithm detects the facial area and applies the detection algorithm over it. The vverall system provides the solution over existing monitoring systems like memory optimization, low power consumption, and detection of sensitive objects. The techniques for object tracking and localization inside the indoor environment were discussed in Sect. 1.2, Chap. 1. Methods to navigate, track, and localize the robot inside the indoor environment are described next in Chap. 5. The practical implementation and challenges are also discussed.

References

1. Ozgul, F., Gok, M., Celik, A., Ozal, Y.: Mining hate crimes to figure out reasons behind. In: 2012 IEEE/ACM International Conference on Advances in Social Networks Analysis and Mining. IEEE, Aug (2012)
2. Feng, M., Zheng, J., Ren, J., Hussain, A., Li, X., Xi, Y., Liu, Q.: Big data analytics and mining for effective visualization and trends forecasting of crime data. IEEE Access 7, 106111–106123 (2019)
3. Yadav, S., Timbadia, M., Yadav, A., Vishwakarma, R., Yadav, N.: Crime pattern detection, analysis & prediction. In: 2017 International Conference of Electronics, Communication and Aerospace Technology (ICECA). IEEE, Apr (2017)
4. Falade, A., Azeta, A., Oni, A., Odun-ayo, I.: Systematic literature review of crime prediction and data mining. Rev. Comput. Eng. Stud. 6(3), 56–63 (2019)
5. Wang, S., Wang, X., Ye, P., Yuan, Y., Liu, S., Wang, F.Y.: Parallel crime scene analysis based on ACP approach. IEEE Trans. Comput. Soc. Syst. 5(1), 244–255 (2018)
6. Jha, G., Ahuja, L., Rana, A.: Criminal behaviour analysis and segmentation using k-means clustering. In: 2020 8th International Conference on Reliability, Infocom Technologies and Optimization (Trends and Future Directions) (ICRITO). IEEE, Jun (2020)
7. Pratibha, Gahalot, A., Uprant, Dhiman, S., Chouhan, L.: Crime prediction and analysis. In: 2nd International Conference on Data, Engineering and Applications (IDEA). IEEE, Feb (2020)
8. Shah, D., Haradi, V.: IoT based biometrics implementation on raspberry pi. Procedia Comput. Sci. 79, 328–336 (2016)
9. Ashby, M.P.J.: The value of CCTV surveillance cameras as an investigative tool: an empirical analysis. Eur. J. Crim. Policy Res. 23(3), 441–459 (2017)
10. Hier, S.P.: Risky spaces and dangerous faces: urban surveillance, social disorder and CCTV. Soc. Legal Stud. 13(4), 541–554 (2004)
11. Ju, J., Ku, B., Kim, D., Song, T., Han, D.K., Ko, H.: Online multi-person tracking for intelligent video surveillance systems. In: 2015 IEEE International Conference on Consumer Electronics (ICCE). IEEE, Jan (2015)
12. Tuan, M.C., Chen, S.L.: Fully pipelined VLSI architecture of a real-time block-based object detector for intelligent video surveillance systems. In: 2015 IEEE/ACIS 14th International Conference on Computer and Information Science (ICIS). IEEE, Jun (2015)
13. Shao, Z., Cai, J., Wang, Z.: Smart monitoring cameras driven intelligent processing to big surveillance video data. IEEE Trans. Big Data 4(1), 105–116 (2018)
14. Prajapati, U., Rawat, A., Deb, D.: Integrated peripheral security system for different areas based on exchange of specific data rates. Wirel. Pers. Commun. 111(3), 1355–1366 (2019)
15. Rawat, A., Deb, D., Rawat, V., Joshi, D.: Methods and systems for data rate based peripheral security (Feb 24 2017), Indian Patent 201721005324 A
16. Prajapati, U., Rawat, A., Deb, D.: A novel approach towards a low cost peripheral security system based on specific data rates. Wirel. Pers. Commun. 99(4), 1625–1637 (2018)
17. Surve, J., Mehta, V., Rawat, A., Kamaliya, K., Deb, D.: Low-cost 2 MHz transmitter for the detection of human trapped under the snow. In: Lecture Notes in Electrical Engineering, pp. 41–50. Springer, Singapore, Nov (2019)
18. Upadhyay, J., Rawat, A., Deb, D., Muresan, V., Unguresan, M.L.: An RSSI-based localization, path planning and computer vision-based decision making robotic system. Electronics 9(8), 1326 (2020)

19. Dollar, P., Wojek, C., Schiele, B., Perona, P.: Pedestrian detection: an evaluation of the state of the art. IEEE Trans. Pattern Anal. Mach. Intell. **34**(4), 743–761 (2012)
20. Loukhaoukha, K., Chouinard, J.Y., Berdai, A.: A secure image encryption algorithm based on Rubik's cube principle. J. Electr. Comput. Eng. **2012**, 1–13 (2012)
21. Upadhyay, J., Deb, D., Rawat, A.: Design of smart door closer system with image classification over WLAN. Wirel. Pers. Commun. (2019)
22. Deb, D., Rawat, A., Khanna, N., Agrawal, S., Radadiya, H.: Smart imperceptible door closing systems with integrated monitoring (2017), Indian Patent 201721021291 A

Chapter 5
Robotic System Configuration with Localization, Path Planning, and Computer Vision

In this chapter, we study the outcomes of simultaneous localization and mapping algorithms from a 3D infrared vision sensor. Also, provided for discussion is the limitation of vision-based and other sensory-based indoor navigation and object tracking methods. At the receiver end, we measure the Received Signal Strength Index (RSSI) of a wireless node which indicates how effectively the device can capture a signal from an access point through a value that effectively indicates whether one has enough signal strength. RSSI is a simple process of location estimation and, when available, ensures that no additional hardware is needed at individual network nodes. It is difficult to estimate the position in an indoor environment directly from the RSSI values.

The robotic system analyzes the environment by capturing the live images. The optical character recognition algorithm identifies characters from images and stores those characters in a string. The algorithm matches the string with the database to get the location information. The results of the predicted position from the classification method are verified with the OCR algorithm. The robotic agent takes the target input from the authorized person. The speech recognition algorithm registers the audio input into a string, and the path optimization algorithm finds the optimal path from the current location to the target. In this chapter, we present a summarized version of the localization, path planning, and computer vision-based Robotic system that is ultimately useful for peripheral security which is the overall theme of this work, and a more detailed formulation is also available in recent literature [1].

In an indoor environment, satellite GPS signals do not have adequate strength due to multi-path reflection from indoor objects [2–6]. A change in data rate is the trigger to sense any movements in a dynamic environment. As the distance increases, the received power (in dB) decreases [7, 8]. The data-rate-based peripheral security system is further implemented and analyzed using a software-defined signal processing software GNU radio.

The power spectral density (dBm/MHz) is also affected by the increment in the distance. The use of a power amplifier reduces the reduction rate of received power concerning the distance [9]. A low-power embedded device can be integrated inside the hydraulic door closer system. Using four transmitter nodes, Naghdi et al. scan and test the area (10.945 m × 6.401 m) under two test conditions with linear movement in the north-south and east-west directions, respectively, to detect human obstacles and correct them as per artificial intelligence algorithms. The result confirmed blockages of more than 87% when using a sliding window received signal strength index (RSSI) method over-sampled inputs. Human presence results in reduced RSSI values during mapping and localization in real time. Classification methods help in correcting such errors [10].

The estimator calculations for the random forest classification model concerning their features provide a better estimation. By calculating the number of estimators, the prediction is faster for real-time data [11]. Computer vision-based indoor navigation uses an approach that utilizes the size, shape, distance from the camera, focusing methods, and different light conditions. Marker-based indoor navigation is deployed [12]. A robotic system should be capable of navigating in an indoor environment and have the capabilities to understand the surroundings. In the visual localization and mapping method, the system builds the 3D point cloud model for position estimation and localization.

5.1 Methodology

The RSSI-based print technique registers multiple sampling inputs from different wireless local area network nodes. The scanning points collecting the signal quality and the signal strength values based on signal-to-noise ratio and data rate are received. The system's pseudo-flow graph is presented in Algorithm 3. A collection of 60, 000 samples per position occurs at the rate of 30 ms/sample through a Raspberry Pi module. While sampling from different locations, the robot also registers the angle and distance for the next and the previous sampling points using a 3-axis magnetometer and an accelerometer. Here, P_s is the respective sampling position, P_e is the estimated position, and DB is the measured RSSI value.

The idle link quality Q is 70, and so the sampling values are taken between 35 and 60. Modules which collect data wait for link quality improvements. Here, n is the number of sampled positions; $r = 1, 2,$ or 3 are the data collected from individual transmitters. θ_{nn} is the angle between corresponding sampling points and D_{nn} is the corresponding distance. P_e is the estimated position calculated from the classification model. As the sample size contains more than 1, 200, 000 samples for 20 nodes (60, 000 per position, that is, 30 ms/sample), the decision-making happens through the prediction and classification model fitness scores.

Algorithm 3: Calculation for position estimation

$P_n => X_{mr}, t_m, Q_{mr}, m = 1, 2, .., > 60, 000$. **Data:** P_s
Result: More than 60,000 sample values per each Position
initialization;
while *Sampling Ends* **do**
   P_s; **if** $35 < q < 65$ **then**
      Sample Q, DB;
      wait until Q increases;
   else
      $Q_{mr} \leftarrow$ varies, update if $Q_{mr} > 55$;
      $DB_{mr} \leftarrow$ RSSI(dB);
      $\theta_{nn} \leftarrow$ Angle of new sampling point;
      $D_{nn} \leftarrow$ Distance between respective sampling point;
   end
end
Data: Applying suitable Prediction model on Real-Time Scan
Result: Real-time Position Prediction
initialization;
while $Q > 60$; *Sample DB* **do**
   $P_s \leftarrow$ Predict Position;
   if $45 < quality < 65$ **then**
     Store P_s
   else
     Predict Again;
   end
   if *(error $= P_s - P_e$);* **then**
     Error < 0.5 m, True Position
   else
     Sample Again
   end
end

Figure 5.1 describes the error nodes in a triangular fashion for a minimal surrounding environment. We record the sample data in a classroom and passage area. The black dotted points were the sample area and Tx_1, Tx_2, and Tx_3 were wireless local area networks. D_1, D_2, and D_3 were the distances between the first sampling points to the individual transmitter nodes. The different locations of wireless node points affect the result. We place the transmitters so that all sampling points get sufficient signal coverage and record the samples in a mesh formation wherein the position numbers are the numbers of the sampling points, and record the data in a zig-zag pattern. A total of 20 samples are collected in 25m² (10 m × 2.5 m) area, and from a constant height of 1 feet distance from the ground surface.

The robot's orientation during the data sampling/mapping time is fixed and set to the North-Zero angle. The system flow graph is shown in Fig. 5.2. We describe all the modules of the system flow graph individually. The system captures the samples with

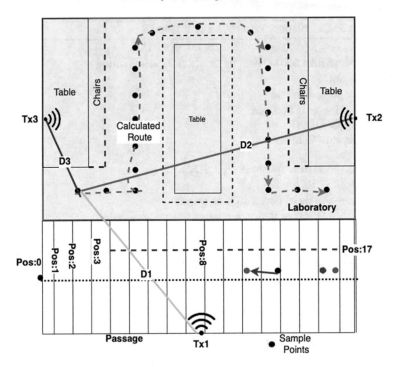

Fig. 5.1 Sampling RSSI values for Indoor positions at a fixed interval of distance

60, 000 samples at a 30 ms/sec sampling rate. At a particular position, the system takes 30 min to collect the samples. The sampling rate is set based on the environmental conditions. The electrical fluctuations and changes in furniture position (or human movements) generate noise in the received signals. The conditions determine the sampling rate that finds the signal possibilities or changes at particular positions. More than 10, 00, 000 samples are taken for the sampled area.

It is difficult to identify the individual positions from collected data due to a lack of uniqueness in their distribution. The distribution pattern is changing concerning the environment dynamics. In this case, the neural network-based classification method can be used as a prediction tool to classify the real-time scan data into an individual position scanned during the area mapping. We can implement several classification methods over the datasets. But the classification method that estimates the result accurately within the range with low detection time is suitable for this proposed method. We consider the path with the minimal hopping points an optimal one because the robot has to verify an accurate location at every point in the course. The system contains the database of the authorized person images. Whenever new and authorized target coordinates are available, the robot follows the new target points and optimizes the new target path.

The generated database gets tested for k-nearest neighbors (KNN), random forest (RF), the radial bias function kernel-support vector machine (RBF SVM), and the

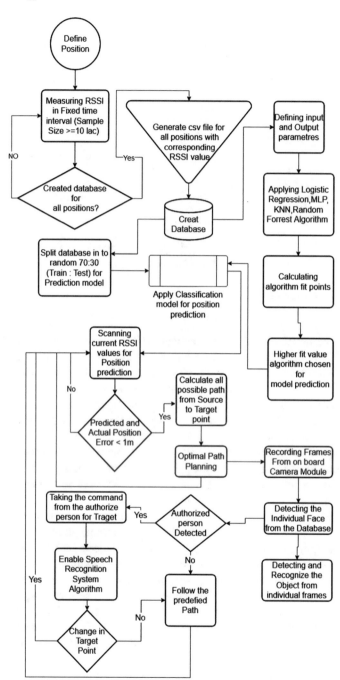

Fig. 5.2   System flow graph

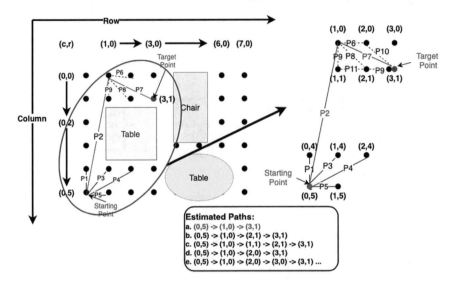

Fig. 5.3 Optimized path planning

multilayer perceptron (MLP) neural network (NN) classification models. We apply the inputs to these models, and we calculate the fit score with RMS error. We then split the database into 70 : 30 training:testing values for prediction. The model gets selected based on their respective higher fit score, minimum root mean square error (RMSE), and processing time. This prediction model then gets applied to the real-time scan values for robot localization. The robot gets scanned for new values and registers the RSSI values only for the link quality between 35 and 70. If the link quality degrades, the prediction model error will increase, indicating the mapped area's wrong robot position. After the position prediction, the algorithm verifies that the calculated position error is within the range, i.e., <1 m or not. After that, the path planning algorithm calculates the optimal path.

At the time of optimal path generation, the path from one fixed position to the target position is decided by connecting the nearest optimal points, as shown in Fig. 5.3. The sampling points are in black dots and the objects in different shapes and colors. We describe the sample node in (column, row) format. If the starting point is (0,5) and the target is (3,1), the optimal path gets shown in red as the number of points connected to nearby reaching points. At the time of sampling, the reaching points are well registered. The starting point (0,5) is connected to the points (0,4), (1,0), (1,4), (2,4), and (1,5). The connecting path from the (1,0) node is a black dotted line. The estimated path from source to destination is calculated by the algorithm and listed in Fig. 5.3.

Figure 5.4 shows the facial and object detection and the associated recognition process.

Individual images are studied to detect the facial area using the Haar cascading method. The processing unit synthesizes the frames captured by a Pi camera and converts them into 128-bit real-valued numbers. A pre-training model generated

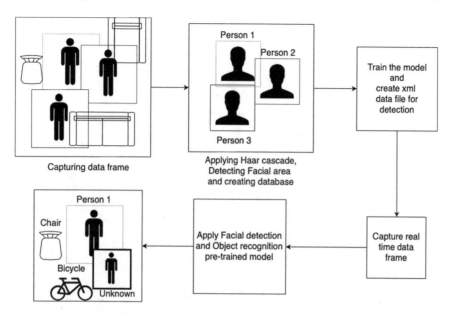

Fig. 5.4 The systematic flow of facial and object recognition algorithm

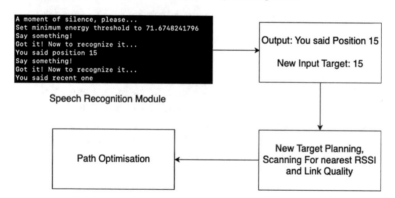

Fig. 5.5 Speech recognition and path planning

based on the Microsoft Common Object in the context dataset was used for object detection of a chair, table, bicycle, etc. This individual detection classifier applied to the resultant image ensures speech recognition for authentic persons. This system takes input in the form of a for a new target. If the system fails to detect and recognize the face, the target point can also be input manually through a remote server. The robotic agent can also take a new target point from the authorized person after the facial detection algorithm approves the person's identity. If the person is authorized then the robotic agent enables the microphone to take the target position at a sampling rate of 2 s. The input stream gets continually registered and converted into a string to remove unnecessary background noise, as shown in Fig. 5.5.

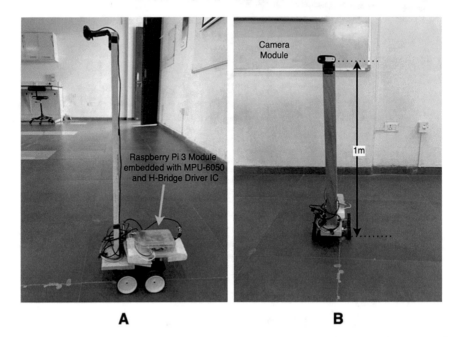

Fig. 5.6 Hardware configuration of a robotic agent: **A** Embedded unit placement **B** Placement of camera module over robotic platform

After getting the new path as input from an authorized person, the system again calculates the optimal route based on the current position. Figure 5.6 shows the hardware setup with the camera setup positioned at 1 m above the ground level and embedded with the Raspberry Pi 3 module to capture an adequately visible area. The proposed robotic system tries to adopt the human cognitive approach to understand and respond to a continually changing environment.

For confirmation or self-verification, the automated system captures the environment's images to analyze them, for instance, frames in an academic institution. For this purpose, with appropriately labeled laboratories and classrooms, the system compares the picture frames with the destination label and location. Figure 5.7 shows such a typically analyzed one. The results are stored in the .text format to compare them with the destination location. The optical character recognition (OCR) algorithm tries to capture all the labels inside the frame. Here the laboratory label is 602, E-YANTRA LABORATORY. Compared with the destination label, if the string matches, then the location is defined as the destination location.

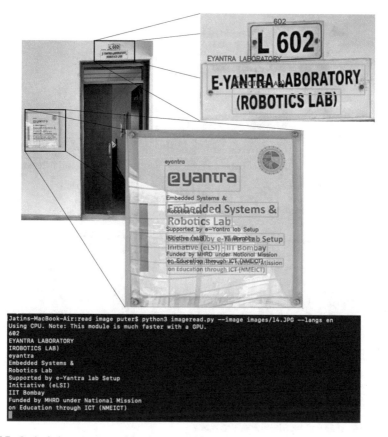

Fig. 5.7 Optical character recognition in captured frames

5.2 Mapping of Collected Samples—Data Distribution and Behavioral Investigation

Figures 5.8, 5.9, 5.10, 5.11, 5.12, 5.13, 5.14, 5.15, 5.16, 5.17, 5.18, 5.19, 5.20, 5.21, 5.22, 5.23, 5.24, 5.25, 5.26, and 5.27 depict the RSSI distribution for individual positions such that Positions 1–20 are the values collected at the time of sampling and localization from three different wireless node points/transmitters (Tx1, Tx2, and Tx3). The number of transmitters required for the specific area and distance between them depends on the range coverage. The individual sampling points can capture the transmitter signal sufficiently. The graphs also show the scanning signal distribution between link quality and the received signal strength index (in dB) for Positions 1–20. The node manufacturer sets the wireless link quality parameters, usually depending upon the signal-to-noise ratio, bit error rate, signal interference plus noise ratio, and packet delivery ratio. The RSSI values vary between (0–70) and (0–100) depending upon the wireless node manufacturer. The signal from the

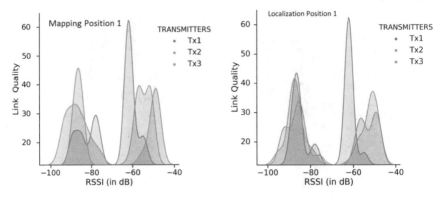

Fig. 5.8 Mapping and localization of Position 1

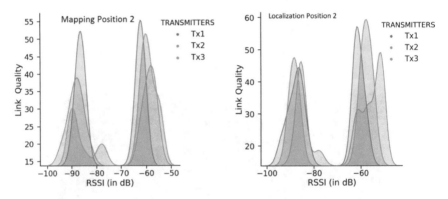

Fig. 5.9 Mapping and localization of Position 2

transmitters have multiple visible spikes and can be seen in Figs. 5.8, 5.10, 5.11, 5.14, 5.17, 5.21, and 5.22 for Positions 1, 3, 4, 7, 10, 14, and 15.

The parameters vary as the position changes. Often, the points nearer to the wireless nodes have higher link quality and RSSI values, as commonly seen in the graphs. As the link quality varies between 0 and 70, it is preferable to wait at the position for improvement in link quality, before registering the data to minimize momentary changes made by any moving object. The collection of data with all possibilities at any position in a dynamic environment leads to a higher accuracy during position prediction. Figure 5.8 shows that Position 1 is near Transmitter 1. The collected parameters show strong spikes. A point away from the transmitter points may see multiple picks at a received signal. Data also shows that there is no consistency in getting strong signal values if a position is near the transmitter. Changing the furniture position or area is affected by external high-voltage line electrical pulses. As shown in Fig. 5.20 (Mapping), Position 13 should get a high signal from Transmitter 2, but the system registers a strong signal from the wireless nodes at Transmitters 3 and 1.

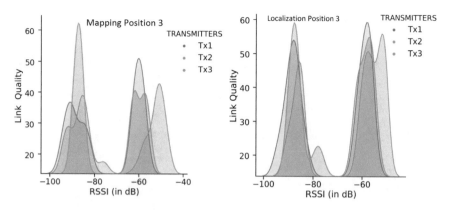

Fig. 5.10 Mapping and localization of Position 3

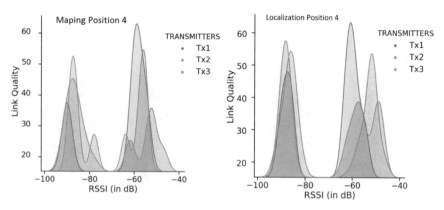

Fig. 5.11 Mapping and localization of Position 4

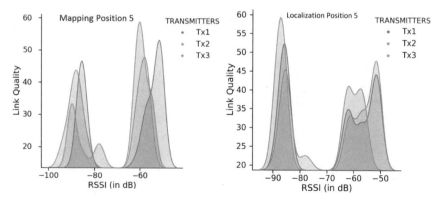

Fig. 5.12 Mapping and localization of Position 5

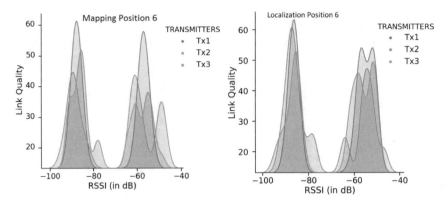

Fig. 5.13 Mapping and localization of Position 6

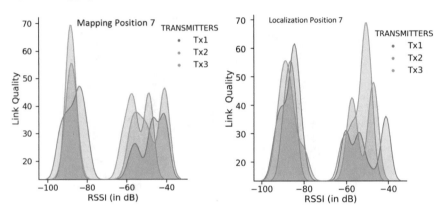

Fig. 5.14 Mapping and localization of Position 7

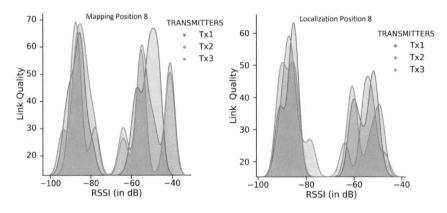

Fig. 5.15 Mapping and localization of Position 8

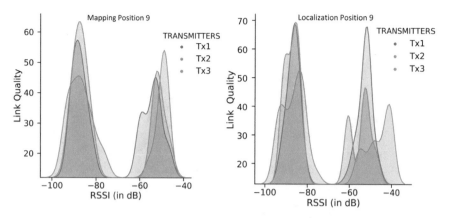

Fig. 5.16 Mapping and localization of Position 9

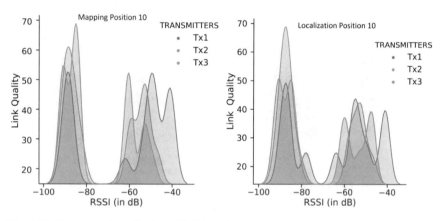

Fig. 5.17 Mapping and localization of Position 10

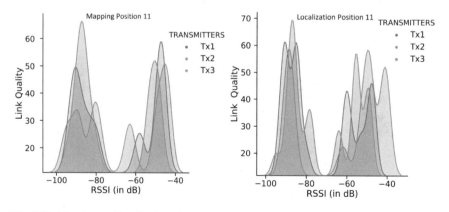

Fig. 5.18 Mapping and localization of Position 11

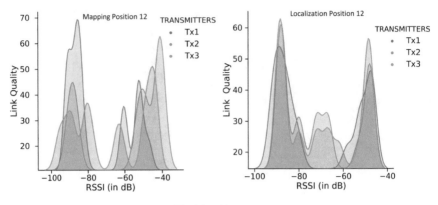

Fig. 5.19 Mapping and localization of Position 12

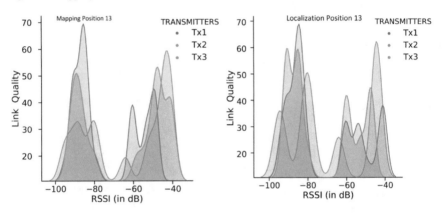

Fig. 5.20 Mapping and localization of Position 13

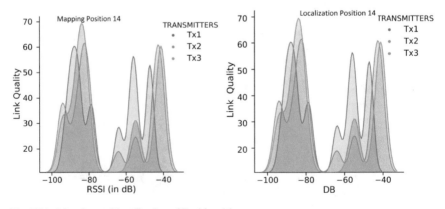

Fig. 5.21 Mapping and localization of Position 14

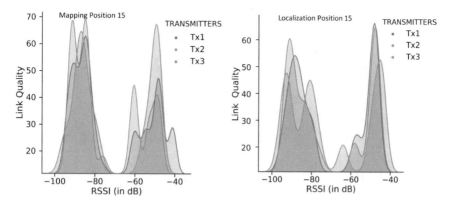

Fig. 5.22 Mapping and localization of Position 15

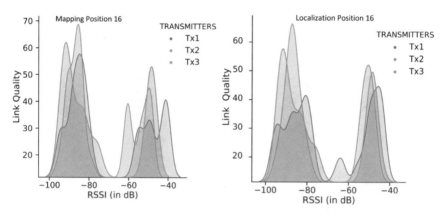

Fig. 5.23 Mapping and localization of Position 16

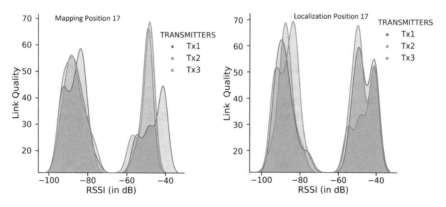

Fig. 5.24 Mapping and localization of Position 17

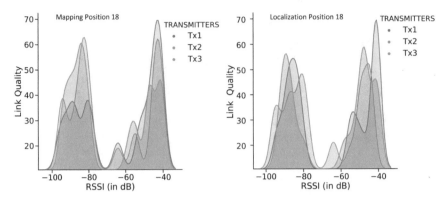

Fig. 5.25 Mapping and localization of Position 18

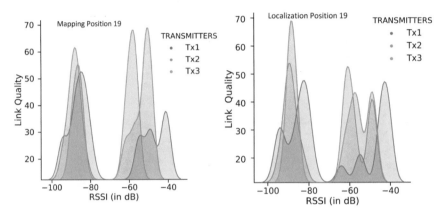

Fig. 5.26 Mapping and localization of Position 19

Figures 5.8, 5.9, 5.10, 5.11, 5.12, 5.13, 5.14, 5.15, 5.16, 5.17, 5.18, 5.19, 5.20, 5.21, 5.22, 5.23, 5.24, 5.25, 5.26, and 5.27 depict the distribution of localization data collected for Positions 1–20. For Position 14 (as seen in Fig. 5.21), the mapping and localization data of Transmitter 3 (green curved region) are matches in many instances. These mapping and localization signal distribution differences are available in many positions. The distribution of Transmitters 1, 2, and 3 are not leading to the distinct identification of a position. The received signal parameters for any position varies with change in the surrounding environment. The overall mapping distribution of individual transmitters nodes for individual positions are shown in Figs. 5.28, 5.29, and 5.30.

The localization distribution for an individual position by distinct wireless nodes is shown in Figs. 5.31, 5.32, and 5.33. The main challenge is to navigate a robotic system in an indoor area using these environment-dependent wireless signals. Direct implementation of statistical methods or classification modules for any sampled dataset may lead to false prediction and position localization. By analyzing the distribution

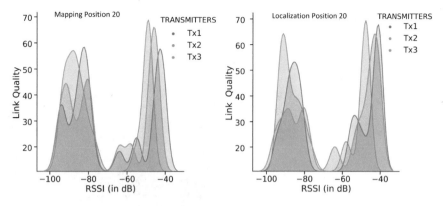

Fig. 5.27 Mapping and localization of Position 20

Fig. 5.28 Wireless
transmitter node 1 mapping
distribution

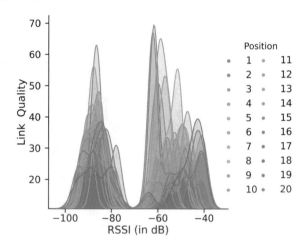

Fig. 5.29 Wireless
transmitter node 2 mapping
distribution

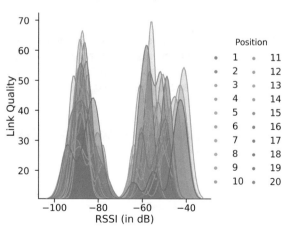

Fig. 5.30 Wireless
transmitter node 3 mapping
distribution

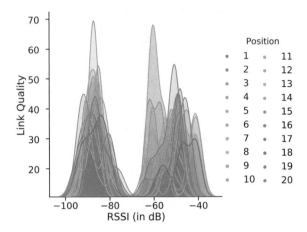

Fig. 5.31 Wireless
transmitter node 1
localization distribution

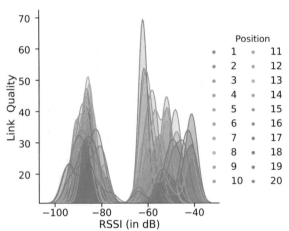

datasets and plots, we conclude that without analysis of the individual classification
accuracy, direct implementation of the prediction model by previous sample trend
cannot assure localization accuracy. It is possible that for each sampled/mapped area,
the classification accuracy may vary.

5.3   Classification Analysis and Results of Image Processing
Models

From the results of the dataset signals given in Figs. 5.18, 5.19, 5.20, 5.21, 5.22,
5.23, 5.24, 5.25, 5.26, and 5.27, the sampled datasets have no unique pattern to dis-
tinguish each position, as is the primary problem of the RSSI-based indoor navigation technique. Neural network-based classification models can give accurate results
compared to other data analysis techniques.

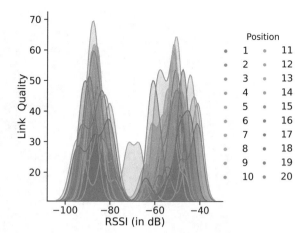

Fig. 5.32 Wireless transmitter node 2 localization distribution

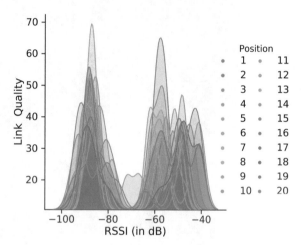

Fig. 5.33 Wireless transmitter node 3 localization distribution

In this proposed method, a different combination of inputs is applied and tested for the same model as shown in Fig. 5.34.

Before the classification process, the steps the model undergoes for training are as follows:

- The training and testing data are split, and the resultant model then gets tested for all the combinations of inputs.
- The input data-set should be shuffled /randomized to make a generalized solution model for the problem.

This condition will reduce variance and make sure that classification models remain generalized for any dataset. This type of classification model is suitable to predict real-time input data. If the input data are not randomized/shuffled, the model may get higher accuracy at the time of testing, but the accuracy significantly reduces for new inputs. Different inputs and the combination are in the first column of Fig.

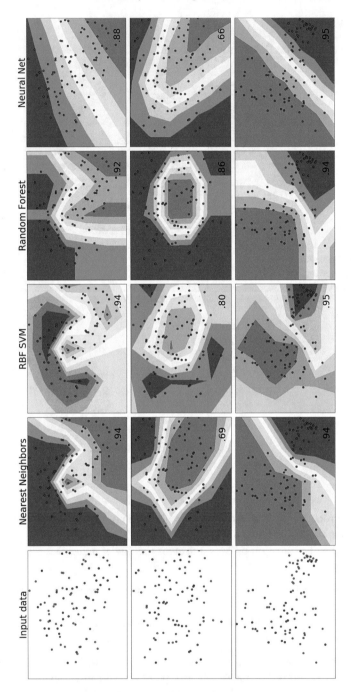

Fig. 5.34 Accuracy analysis of different classification methods

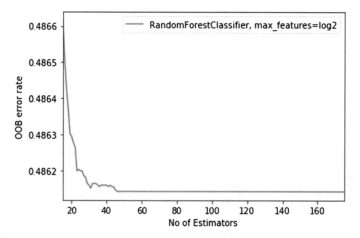

Fig. 5.35 Number of minimum estimators require for limiting OOB error

5.34. The performance of the individual model and its accuracy varies for different set combinations. The point of model comparison is to illustrate the characteristics of the decision boundary of an individual classifier. The selection of the final classification model is important and relates to the overall navigation accuracy. Figure 5.34 shows the comparative results among the best performing classification models. The inputs are the link quality and RSSI values, and positions are the outputs. The randomized data segregates into 70:30 training:testing datasets. The solid red dots are the training data, and the light-colored dots are the testing data. The red and blue regions are the boundary conditions to separate points. The performance analysis of different classification models is done to calculate the mean absolute error (MAE), mean square error (MSE), and root mean square error (RMSE) for sampled and real-time scanned datasets. The RF classification method provides lower values of MAE_{DB}, MSE_{DB}, MAE_{RT}, MSE_{RT}, and $RMSE_{RT}$ compared to other methods. The RF classifier is computationally the fastest for sampled and real-time scanned datasets.

Figure 5.35 shows that the number of estimators requires one to achieve a lower out-of-bag (OOB) error rate of <50 for the sampled dataset.

The estimator ranges from 15 to 175. However, considering the estimator range higher than 50 only increases the processing time without having any impact on the OOB error rate. The error rate begins to drop only when the number of estimators is substantially lower than 50. The higher the RSSI value, the higher the signal strength.

Figure 5.36 shows the bar graph of actual and predicted values for 25 random cases of 20 positions. The sample positions are on the Y-axis, and the random samples taken from the prediction results are on the X-axis. The robot performs position prediction using the real-time scan data and prediction model. If for a particular point, there is an available precise prediction as per the assumptions made by a robot, then before moving, the robot records that point as an accurate estimation. The cumulative

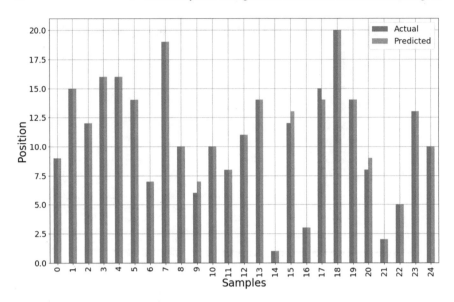

Fig. 5.36 Visualization of actual and predicted values for individual sampled positions

distribution of errors (in meters) is determined. Overall data distribution is shown on the scale of 0–1, with the error (in meters) on the X-axis.

In an average sense, the RF classification method gets its 90% position predictions within the 3 m range. The distance between the transmitters is set such that all the points receive proper coverage from different transmitters. Additionally, the RF-based method gets a maximum error of 6 m for any position. The RF, KNN, RBF SVM, and MLP achieve 90% accuracy at 2.87, 3.7, 3.85, and 5 m, as shown by vertical dotted lines in Fig. 5.37. The proposed system is implemented in an indoor area with modification in transmitter placement, scaling in wireless nodes, and sample points. The accuracy will be the same or can increase if the samples get collected in diversified conditions. The object detection and recognition module and the facial detection and recognition module work independently. Object detection results are shown in Fig. 5.38.

The recognition of a detected person happens through a respective database name. A person who has no record in the database is an unknown individual, as shown in Fig. 5.39. The face orientation and lighting conditions may affect the detection process, and so the robot tries to orient differently for appropriate adjustment to the lighting conditions. The OCR algorithm is the advanced step of cognition for a robotic system. The robot confirms the location and analyzes the environment just as a human being would. The object and facial detection and recognition with OCR capabilities help the robot to take appropriate decisions.

For the sampled dataset, the random forest classification method provides a promising result with the least MAE, MSE, and RMS error compared to other prediction models. The path optimization algorithm chooses the lowest hopping node path

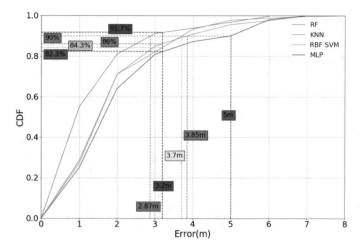

Fig. 5.37   Representation of position error using cumulative distribution function

Fig. 5.38 Output of object
detection and recognition
algorithm

as the shortest path. In the proposed method, the RF classification method achieved a
location accuracy of 5.7, 7.4, and 9.4% higher than RBF SVM, KNN MLP in a 3.2 m
span. The designed robotic platform can be further modified into transport laboratory
equipment from one room to another on the same floor. The system can sense the
surroundings and make immediate decisions using the proposed computer vision.
In future, it is possible to deploy this system on aerial vehicles inside buildings for
surveillance purposes.

Navigating and localizing a closed building or indoor environment is challenging, as seen in received signal distribution during localization and mapping.
The researchers are continuously working to achieve centimeter-level accuracies.
The wireless node-based localization approach is one technique to map and localize
the area with pre-installed wireless nodes. This chapter concluded that all of the

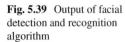

Fig. 5.39 Output of facial
detection and recognition
algorithm

classification techniques should be tested and compared over the collected sampled
data before selecting the prediction algorithm. The classification methods used in
any other installation site can have different accuracy levels in other sampling areas.
The different classification methods are analyzed in samples collected for this study.
Based on the results, the RF-based classification method has a lower prediction time
with higher accuracy than other dynamic environments. The robotic system navigates
inside the indoor environment with the help of an RSSI-based indoor navigation algorithm and computer vision-based verification methods. The technology can predict
the fault in the implemented system before the situation goes out of the control state.
A railway track line failure can be detected by comparing the signal patterns of different load-based track vibrations. The track vibration is measured during the train
passing through the track and when minor track cracks are present. The methodology
is described in Chap. 6.

References

1. Upadhyay, J., Rawat, A., Deb, D., Muresan, V., Unguresan, M.L.: An RSSI-based localization,
   path planning and computer vision-based decision making robotic system. Electronics **9**(8),
   1326 (2020)
2. Sithole, G., Zlatanova, S.: Position, location, place and area: an indoor perspective. ISPRS
   Ann. Photogramm. Remote Sens. Spat. Inf. Sci. III-4, 89–96 (2016)
3. Errington, A.F.C., Daku, B.L.F., Prugger, A.F.: Initial position estimation using RFID tags: a
   least-squares approach. IEEE Trans. Instrum. Meas. **59**(11), 2863–2869 (2010)
4. Tesoriero, R., Gallud, J., Lozano, M., Penichet, V.R.: Tracking autonomous entities using rfid
   technology. IEEE Trans. Consum. Electron. **55**(2), 650–655 (2009)
5. Saab, S.S., Nakad, Z.S.: A standalone RFID indoor positioning system using passive tags.
   IEEE Trans. Ind. Electron. **58**(5), 1961–1970 (2011)
6. Kjærgaard, M.B., Blunck, H., Godsk, T., Toftkjær, T., Christensen, D.L., Grønbæk, K.: Indoor
   positioning using GPS revisited. In: Lecture Notes in Computer Science, pp. 38–56. Springer,
   Berlin, Heidelberg (2010)

7. Prajapati, U., Rawat, A., Deb, D.: A novel approach towards a low cost peripheral security system based on specific data rates. Wirel. Pers. Commun. **99**(4), 1625–1637 (2018)

8. Rawat, A., Deb, D., Rawat, V., Joshi, D.: Methods and systems for data rate based peripheral security (Feb 24 2017), Indian Patent 201721005324 A

9. Prajapati, U., Rawat, A., Deb, D.: Integrated peripheral security system for different areas based on exchange of specific data rates. Wirel. Pers. Commun. **111**(3), 1355–1366 (2019)

10. Naghdi, S., O'Keefe, K.: Detecting and correcting for human obstacles in BLE trilateration using artificial intelligence. Sensors **20**(5), 1350 (2020)

11. Hastie, T., Tibshirani, R., Friedman, J.: The Elements of Statistical Learning. Springer, New York (2009)

12. Delfa, G.C.L., Catania, V., Monteleone, S., Paz, J.F.D., Bajo, J.: Computer vision based indoor navigation: a visual markers evaluation. In: Ambient Intelligence - Software and Applications, pp. 165–173. Springer International Publishing (2015)

Chapter 6
Secured System Design for Failure Mitigation of Railway Tracks

This chapter proposes methods for failure mitigation of railway tracks using a predefined strain gauge module-based network to secure railway networks. This module's primary purpose is to analyze the real-time condition monitoring of railway tracks to prevent railway accidents due to track crack/deformation, caving of support strata underneath, and various other reasons. Failure may be because of the track's life cycle, natural disasters, or intentionally some anti-social elements damaging the rail track [1]. In this method, we measure and record changes in the railway track load and vibrations to enable the prediction of fatal accidents due to derailment and prevent the loss of thousands of lives and other valuable resources. It detects the instantaneous change in the track's response, records at various levels where it is read, analyzed, and takes corrective measures on the faulty portion of the railway track depending upon its severity level. With the new era, there are several enhancements needed in the devices used for the security of the train and its passengers. Due to many reasons like improper maintenance of tracks, inefficient track parts, improper switching of tracks, water runoff below tracks, etc., rail accidents are happening and are causing harm to humans as well as other resources [2, 3].

The Indian Railways has in the recent past introduced a concept Ultrasonic Flaw Detection (USFD) testing of rails, in which the track is checked in a scheduled fashion after the passage of 8 gross million tonnes of traffic [4–6]. The present competency of manual testing is a time-taking and resource-exhausting one. A high degree of operator skills and integrity is needed for its task. Hence, the need for trained and certified personnel is more. Moreover, no permanent record of the inspection data is there for evaluation by using the ultrasonic method. Spurious indications, and the misreading of signals, can lead to unnecessary repairs, which is why a validated process should be practiced [7–10].

Nowadays, pneumatic systems for brakes are also being used for avoiding derailment to much extent. But such systems are cumbersome and require human handling.

© The Author(s), under exclusive license to Springer Nature Singapore Pte Ltd. 2021 97
A. Rawat et al., *Recent Trends In Peripheral Security Systems*, Services and Business Process Reengineering, https://doi.org/10.1007/978-981-16-1205-3_6

Due to human involvement, chances of error becomes more. It works after the derailment occurs, so it is not helpful for anticipating the faults prior to their occurrence. The proposed method using a predefined strain gauge module-based network provides accurate information about tracks, to avoid such accidents in order to cause the least harm. Existing devices provide safety at a very conventional level, such that in order to yield good outcomes this module is incorporated. It leads to a system having considerable good efficiency with reliable results, along with the data analysis done by a module consisting of a specific strain gauge with an RF transmitter network which will transfer the data to the respective personnel for further evaluation. Such analyzed data are helpful in designing new track parts by taking care of the bounds of parameters at the industrial level. This method enables us to monitor the conditions of railway tracks without being around them. Remotely placed tracks are also inspected continuously, and relevant measures are taken if any fault occurs [11–13].

6.1   Design of Predefined Strain Gauge Module-Based Network

The proposed system consists of three sub-systems which are Data Acquisition and Transmission Module (DATM), Data Reception Module (DRM), and Base Station (BS). The DATM has basic modules such as (i) Strain Gauges and (ii) Transmitter, whereas DRM has (i) Receiver, (ii) Microcontroller, and (iii) Transmitter. Base Station (BS) has (i) Receiver, (ii) Monitoring System, and (iii) Data Analysis and Interpretation. Data acquisition and Transmission Module (DATM) detects the changes in the "n-motion weight and vibration parameters" by use of strain gauges. Strain Gauge output is sensitive to every change in the weight as well as vibration parameters. Such parameters are sent to the DRM through a Transmitter. Power input to this module is very less which enables it to provide an inexpensive approach for persistent condition monitoring of track parameters.

Data Acquisition and Transmission Module (DATM) is arranged in a different manner depending upon different track conditions like high terrains, along railroad switches, etc. It is arranged in different unique patterns depending upon the surrounding ecosystem. The block diagram of the DATM system and arrangement is shown in Figs. 6.1 and 6.2 which contain the load cell and the transmitter unit. This module is resilient enough to withstand alterations in its neighborhood provisions.

Fig. 6.1 Components of DATM system

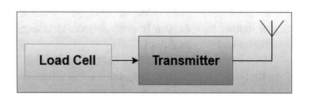

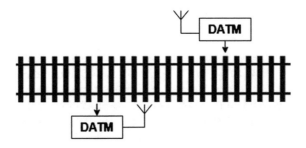

Fig. 6.2 DATM module placement of track

Fig. 6.3 Block diagram of DRM

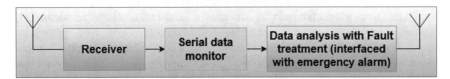

Fig. 6.4 Block Diagram of Base Station (BS)

The DRM module is linked with the microcontroller module for better control characteristics and multiplexing capability of the signal of different DATM. The process is shown in Fig. 6.3.

Microcontroller synchronizes the operation of all components of DRM. DRM transmits the signals received from DATM to Base Station (BS) which de-multiplex the signal and further deeply analyze the signal to perfectly monitor the condition of tracks as shown in Fig. 6.4. Transmitted track parameter data are continuously sent to Data Reception Module (DRM) which receives the data in order to take corrective measures accordingly. Such live data are monitored with minimum errors and more precision at the base station. The base station is located at the nearest railway station which receives and handles the data in order to provide regulated protection. This invention is capable of preventing sudden as well as incipient faults. Due to its distinguished way of monitoring the data and handling it, this invention has the caliber of acting upon the faults as per its severity. It gives a non-conventional level approach for the safety of the railway network [14].

6.2   Operational Procedure

In this method, the process starts with a strain gauge which is kept at regular intervals under the railway track. At a given time, the instant location of the train and that of the strain gauge station is known in prior, enabling us to predict the standard response from DATM. This standard response is defined over a narrow band referred to as a safe response zone. The safe response zone is defined as per the threshold level of the track parameters. Estimation of the safe response zone is done due to consideration of dynamic load transferred, frequency of load repetition, boundary parameters, and so on; any response outside of the safe response zone will then be flagged by the analysis tool as deviation from normal and will raise an alarm bell sounding for attention from the system operation. Depending on the severity of the deviation, the system operator can then instruct the necessary precautions [15, 16].

The strain gauge is capable of producing the proportional signals as per the variations in the load and vibrational parameters along the track horizontally as well as vertically. So signals produced by the strain gauge are transmitted with the help of a Radio Frequency (RF) Transmitter. Strain gauge and RF transmitter both are included in Data Acquisition and Transmission Module (DATM). The strain gauge is kept at various positions across the Railway Tracks. It generates the voltage proportional to the weight and vibrational changes along it. It is used for consistently getting the information about the load and vibrational changes done along the Tracks. It is then connected to the transmitter which modulates, amplifies, and filters the produced signals. Such signals are transmitted at a particular frequency through an antenna to the nearest Data Reception Module (DRM).

DATM modules are connected in an efficient manner along the railway tracks situated at the different terrains. These modules send the data to the nearby Data Reception module (DRM) individually as per the frequency assigned to each signal sent to the respective DRM. DATM modules are kept after a particular fixed distance in an alternate pattern, in order to collect the data at different places at the same time. These modules are also kept along the shifter tracks to obtain the variation in the load data. Such modules are properly calibrated and tuned for the proper functioning of the overall system. The DRM is the second module after DATM, which is a subsequent step after the real-time data from the railway track. DRM receives data from various DATM via antenna as per the frequency tuning. All the data which is collected by the receiver is dispatched to the microcontroller of DRM. In the microcontroller, the data generated by the strain gauge of DATM is enumerated and multiplexed exhaustively. After computing data of each and every nearby DATM, enumerated data is a catapult to the transmitter of the DRM. The transmitter of the DRM sends all the data through the antenna to the vicinal base station. This computed data of the strain gauge will be received at the base station.

The base station is the successive and the final module after DRM in this invention. Detailed calculated data from all nearby DRM is received at the base station with the help of an antenna and then de-multiplexed. The receiver of the base station fetches the data from all nearby DRM modules. All the data from the receiver of the base station is directed toward the system where continuous monitoring of such data is done. So, the signal generated by the strain gauge kept under the railway track goes through a rigorous process so that no single fault situation is ignored till the monitoring of this data is done in detail. After so much of data analyzing, the system is able to differentiate the faulty signal from the normal signals. This faulty signal is checked and as per the severity, appropriate actions are taken accordingly. In case of any emergency, the alarm is activated which resolves the fault conditionally. Or else, if there is no fault on the railway track, all this data is recorded which can be later used to check the wear and tear of the railway track. So, in this manner without being physically present around the railway track, the authority will be able to know where maintenance is needed. It also provides the system for periodic Inspection of remote railway tracks. The data analyzed is to be used for the construction of track components considering the safety of the resources as the primary priority. The variation of load and vibration parameters of the track will be different for the faulty/broken track with a loose base (due to rain).

So basically the operation of this system starts with the DATM. The main component of DATM is the strain gauge, which senses the load and vibrations applied to it and generates proportional signals. These signals from all DATM are sent to the nearest DRM at a particular frequency. DRM is the second module of this invention. DRM receives data from all surrounding DATM, which are then classified and processed in the microcontroller of the DRM. Processed data from DRM is sent to the neighboring base station. Here, data is continuously monitored by which the system will detect any faulty condition so that appropriate actions for the failure mitigation will be taken and derailment of the train will be prevented. Initially, the base station applying the D-MUX operation and classifying the load and vibration data depend on their location. After that, it will remove the noise and check whether it crosses the predefined threshold level of load/vibration of the track. Depending upon that, it will decide the possible condition of the track. The threshold level of load/vibration is predefined by the various experiments of different tracks of the different possible fault and normal conditions.

6.3 Experimental Setup and Result Discussion

A small laboratory working model was prepared with the help of a toy train. The experimental setup had a vibration sensor, Arduino, and toy train. The major reason behind derailment is the malfunctioning of the railway track or missing fishplate. So the experiment was mostly concentrated on the railway track of our toy train. Capturing the data of track vibration was the pivotal work. For that, vibration sensor SW-420 with LM393 comparator was used. In the experimental setup, we mainly focused on two states, normal state and faulty state. In the normal state, the railway track of the toy train was totally connected with no faults in the train and the track. Vibration data of the track was collected when the train was made to run on the track. In the faulty state, part of the track was kept open loosely. On such defected railway

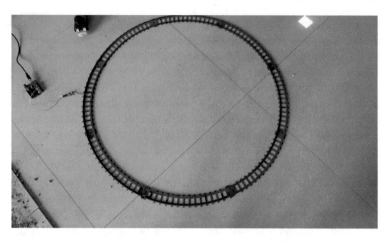

Fig. 6.5 Normal state of toy train track

Fig. 6.6 Faulty state of toy train track

track of the toy train, vibration data was taken. This was done to differentiate between the vibration of normal and faulty states. Normal and faulty states are shown in Figs. 6.5 and 6.6, respectively. But in the faulty connection of the track, heavy frequency fluctuation was noticed and that can be seen in Fig. 6.7.

Real-time data is taken from the railway track after a successful in-house experimental results. The sensor was placed on the foot of the rail lines around 2 Km away from the station, so that the vibration data collected can be rich in data points and data processing can be easily done. Data collected in the main work was to extract the main features of railway track vibration. From this, it can be said as to which parameter the vibration of the railway track is dependent on. Classification of data was done on various parameters like peak amplitude, frequency contents, and period of vibration. Since testing is still in progress, we are presenting only a few details.

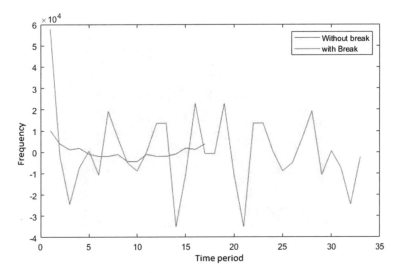

Fig. 6.7 Comparing normal and faulty states

Fig. 6.8 On-field data
acquisition setup

The real-time vibration sensing system setup with a railway track is shown in Fig.
6.8 and the vibration sensing module is shown in Fig. 6.9.

Comparison of the frequency content of a few trains under test is shown in Fig.
6.10.

From the experiments performed on the toy train and a few actual trains, it is
observed that the state of the railroad track can be effectively checked using vibration

Fig. 6.9 Setup with memory card

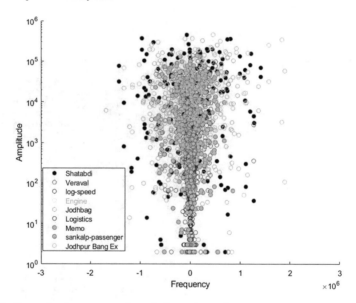

Fig. 6.10 Comparing frequency content of a few trains under test

investigation of the track. With more information procurement and examination, a general equation can be characterized to give a vibration threshold of an explicit train. Anything above it will ring an alarm. Subsequently, we can anticipate any future setbacks because of the railroad track fault. By creating a general equation by collecting other various types of data, it can be made sure that vibration from the track is faulty. Using this module, we can inspect the tracks at remote locations very keenly without much manual effort. From the practical implementation, it is seen that the vibration signal patterns for a normal and a faulty track line are different.

The signal pattern and distribution vary based on the load on the track. The pattern of faulty track signals can be predicted using different neural network-based regression and classification techniques. We can implement neural network-based analysis after collecting a large sample of data in a specific time period.

References

1. Toliyat, H., Abbaszadeh, K., Rahimian, M., Olson, L.: Rail defect diagnosis using wavelet packet decomposition. IEEE Trans. Ind. Appl. **39**(5), 1454–1461 (2003)
2. Timusk, M., Lipsett, M., Mechefske, C.K.: Fault detection using transient machine signals. Mech. Syst. Signal Process. **22**(7), 1724–1749 (2008)
3. Lad, P., Pawar, M.: Evolution of railway track crack detection system. In: 2016 2nd IEEE International Symposium on Robotics and Manufacturing Automation (ROMA). IEEE, Sep (2016)
4. Lombaert, G., Degrande, G., François, S., Thompson, D.J.: Ground-borne vibration due to railway traffic: a review of excitation mechanisms, prediction methods and mitigation measures. In: Notes on Numerical Fluid Mechanics and Multidisciplinary Design, pp. 253–287. Springer, Berlin, Heidelberg (2015)
5. Ono, K., Yamada, M.: Analysis of railway track vibration. J. Sound Vib. **130**(2), 269–297 (1989)
6. Aher, S.B., Tiwari, D.: Railway disasters in India: causes, effects and management **6**, 125–132 (2019)
7. Doebling, S.W., Farrar, C.R., Prime, M.B.: A summary review of vibration-based damage identification methods. Shock Vib. Dig. **30**(2), 91–105 (1998)
8. Firlik, B., Czechyra, B., Chudzikiewicz, A.: Condition monitoring system for light rail vehicle and track. Key Eng. Mater. **518**, 66–75 (2012)
9. Yang, M., Edge, K.A., Johnston, D.N.: Condition monitoring and fault diagnosis for vane pumps using flow ripple measurement
10. Clark, R.: Rail flaw detection: overview and needs for future developments. NDT & E Int. **37**(2), 111–118 (2004)
11. Papaelias, M.P., Roberts, C., Davis, C.L.: A review on non-destructive evaluation of rails: state-of-the-art and future development. Proc. Inst. Mech. Eng. Part F J. Rail Rapid Transit **222**(4), 367–384 (2008)
12. Timusk, M.A.: A unified method for anomaly detection in unsteady systems. ProQuest (2006)
13. Cannon, D.F., Edel, K.O., Grassie, S.L., Sawley, K.: Rail defects: an overview. Fatigue Fract. Eng. Mater. Struct. **26**(10), 865–886 (2003)
14. Mungule, M., Rawat, A., Khanna, N., Mishra, S., Mishra, P.: Method and system for failure mitigation of railway tracks using predefined strain gauge module based network (Oct, 25 2017), Indian Patent 01721037691 A
15. Kumar, K.A., Reddy, D.M.: Application of frequency response curvature method for damage detection in beam and plate like structures. IOP Conference Series: Materials Science and Engineering, vol. 149, p. 012160, Sep (2016)
16. Lin, H., Liu, X., Ren, J., Shubber, A.A.M.: Dynamic responses of ballastless track with damaged cracks. In: International Conference on Transportation Engineering 2009. American Society of Civil Engineers, Jul (2009)

Epilogue

An integral approach is ensured for an overall security system by interfacing change in data-rate-based tracking system in an outdoor environment with the integration of computer vision-based enhanced security system for indoors. The robotic agent can be deployed inside the indoor environment to transfer materials between two destination points and area patrolling. The agent can also be deployed in the area of restricted GPS signals. The image processing target tracking-based object localization and formation-based drones can be deployed in the GPS prone area. The formation can be made so that the object is always visible in any of the tracking drones. The failure of railway tracks is due to minor cracks formed due to environmental temperature changes or the lack of scheduled maintenance. The system can further be modified to sense the life span of the track material after frequent use and the type of transportation placed over it. The system can transfer the signals to the control center to inspect the track area based on the previous data pattern. This can save the life of any person.

The RF-based tracking system locates an object or person that is trapped under the snow. This system can be integrated with the data-rate-based security system to track and identify an object that is not visually seen by the camera modules. Implementation of computer vision-based object tracking for thermal images can be the next stage of progression. A track line fault-based system can be used to find out the unauthorized movement in a specific area by predicting the ground's vibration signal and the weight of the moving objects.

Individual techniques proposed in this book can be implemented with a single command and the control center unit. Integrating each system minimizes the independent system limitation and expands the coverage of individuals. Also, the hardware implementation of tracking algorithms is minimal and can be more affordable after mass production. The system implantation difficulty is lower than other state-of-the-art methods by understanding the practical implementation difficulties. The practical installation errors are filtered by the robust algorithm design. The proposed system design application is limited to a specific organization and can be used for a financial or education institution to intensify the level of protection.

A. Rawat et al., *Recent Trends In Peripheral Security Systems*, Services and Business Process Reengineering, https://doi.org/10.1007/978-981-16-1205-3

Printed in the United States
by Baker & Taylor Publisher Services